ISSN 1534-1585

ENERGY: SUPPLIES, SUSTAINABILITY, AND COSTS

Kim Masters Evans

INFORMATION PLUS® REFERENCE SERIES
Formerly Published by Information Plus, Wylie, Texas

GALE
CENGAGE Learning·

Detroit • New York • San Francisco • New Haven, Conn • Waterville, Maine • London

GALE
CENGAGE Learning·

Energy: Supplies, Sustainability, and Costs

Kim Masters Evans

Kepos Media, Inc.: Paula Kepos and Janice Jorgensen, Series Editors

Project Editors: Kathleen J. Edgar, Elizabeth Manar, Kimberley McGrath

Rights Acquisition and Management: Margaret Chamberlain-Gaston

Composition: Evi Abou-El-Seoud, Mary Beth Trimper

Manufacturing: Rita Wimberley

Cover photograph: © Greg Randles/Shutterstock.com.

Gale
27500 Drake Rd.
Farmington Hills, MI 48331-3535

ISBN-13: 978-0-7876-5103-9 (set)   ISBN-10: 0-7876-5103-6 (set)
ISBN-13: 978-1-4144-8139-5   ISBN-10: 1-4144-8139-X

ISSN 1534-1585

This title is also available as an e-book.
ISBN-13: 978-1-5730-2285-9 (set)
ISBN-10: 1-5730-2285-3 (set)
Contact your Gale sales representative for ordering information.

Printed in the United States of America
1 2 3 4 5   17 16 15 14 13

TABLE OF CONTENTS

PREFACE

Energy: Supplies, Sustainability, and Costs is part of the *Information Plus Reference Series*. The purpose of each volume of the series is to present the latest facts on a topic of pressing concern in modern American life. These topics include today's most controversial and studied social issues: abortion, capital punishment, care for the elderly, child abuse, crime, the environment, health care, immigration, minorities, national security, social welfare, women, youth, and many more. Even though this series is written especially for high school and undergraduate students, it is an excellent resource for anyone in need of factual information on current affairs.

By presenting the facts, it is the intention of Gale, Cengage Learning to provide its readers with everything they need to reach an informed opinion on current issues. To that end, there is a particular emphasis in this series on the presentation of scientific studies, surveys, and statistics. These data are generally presented in the form of tables, charts, and other graphics placed within the text of each book. Every graphic is directly referred to and carefully explained in the text. The source of each graphic is presented within the graphic itself. The data used in these graphics are drawn from the most reputable and reliable sources, such as from the various branches of the U.S. government and from private organizations and associations. Every effort was made to secure the most recent information available. Readers should bear in mind that many major studies take years to conduct, and that additional years often pass before the data from these studies are made available to the public. Therefore, in many cases the most recent information available in 2013 is dated from 2010 or 2011. Older statistics are sometimes presented as well, if they are landmark studies or of particular interest and no more-recent information exists.

Although statistics are a major focus of the *Information Plus Reference Series*, they are by no means its only content. Each book also presents the widely held posi-tions and important ideas that shape how the book's subject is discussed in the United States. These positions are explained in detail and, where possible, in the words of their proponents. Some of the other material to be found in these books includes historical background, descriptions of major events related to the subject, rele-vant laws and court cases, and examples of how these issues play out in American life. Some books also feature primary documents or have pro and con debate sections that provide the words and opinions of prominent Amer-icans on both sides of a controversial topic. All material is presented in an evenhanded and unbiased manner; readers will never be encouraged to accept one view of an issue over another.

HOW TO USE THIS BOOK

The United States is the world's largest consumer of energy in all its forms. Gasoline and other fossil fuels power its cars, trucks, trains, and aircraft. Electricity generated by burning oil, coal, and natural gas—or from nuclear or hydroelectric plants—runs Americans' lights, telephones, televisions, computers, and appliances. With-out a steady, affordable, and massive amount of energy, modern America could not exist. This book presents the latest information on U.S. energy consumption and pro-duction and compares it with years past. Controversial issues such as U.S. dependence on foreign oil and gov-ernment subsidies for the fossil fuel industries are explored.

Energy: Supplies, Sustainability, and Costs consists of nine chapters and three appendixes. Each of the major elements of the U.S. energy system—such as coal, nuclear energy, renewable energy sources, and electricity generation—has a chapter devoted to it. For a summary of the information that is covered in each chapter, please see the synopses that are provided in the Table of Con-tents. Chapters generally begin with an overview of the basic facts and background information on the chapter's topic, then proceed to examine subtopics of particular interest. For example, Chapter 8: Electricity explains how electricity is generated, transmitted, and distributed to consumers. Information is provided about the various fuel sources (coal, natural gas, nuclear energy, hydroelectric power, and other renewable sources) that are used to produce electricity and the economic sectors (residential, commercial, industrial, and transportation) that consume electricity. Trends in electricity prices are described, and market factors that affect pricing are discussed. This chapter also supplies projections prepared by the U.S. Department of Energy's Energy Information Administration regarding future domestic and international electricity generation. Finally, various environmental issues that are associated with the electric power sector are briefly summarized. Readers can find their way through a chapter by looking for the section and subsection headings, which are clearly set off from the text. They can also refer to the book's extensive Index, if they already know what they are looking for.

Statistical Information

The tables and figures featured throughout *Energy: Supplies, Sustainability, and Costs* will be of particular use to readers in learning about this topic. These tables and figures represent an extensive collection of the most recent and valuable statistics on energy production and consumption—for example, the amount of coal mined in the United States in a year, the rate at which energy consumption is increasing in the United States, and the percentage of U.S. energy that comes from renewable sources. Gale, Cengage Learning believes that making this information available to readers is the most important way to fulfill the goal of this book: to help readers understand the topic of energy and reach their own conclusions about controversial issues related to energy use and conservation in the United States.

Each table or figure has a unique identifier appearing above it, for ease of identification and reference. Titles for the tables and figures explain their purpose. At the end of each table or figure, the original source of the data is provided.

To help readers understand these often complicated statistics, all tables and figures are explained in the text. References in the text direct readers to the relevant statistics. Furthermore, the contents of all tables and figures are fully indexed. Please see the opening section of the Index at the back of this volume for a description of how to find tables and figures within it.

Appendixes

Besides the main body text and images, *Energy: Supplies, Sustainability, and Costs* has three appendixes. The first is the Important Names and Addresses directory. Here, readers will find contact information for a number of organizations that study energy. The second appendix is the Resources section, which is provided to assist readers in conducting their own research. In this section, the author and editors of *Energy: Supplies, Sustainability, and Costs* describe some of the sources that were most useful during the compilation of this book. The final appendix is the Index. It has been greatly expanded from previous editions and should make it even easier to find specific topics in this book.

ADVISORY BOARD CONTRIBUTIONS

The staff of Information Plus would like to extend its heartfelt appreciation to the Information Plus Advisory Board. This dedicated group of media professionals provides feedback on the series on an ongoing basis. Their comments allow the editorial staff who work on the project to continually make the series better and more user-friendly. The staff's top priority is to produce the highest-quality and most useful books possible, and the Information Plus Advisory Board's contributions to this process are invaluable.

The members of the Information Plus Advisory Board are:

- Kathleen R. Bonn, Librarian, Newbury Park High School, Newbury Park, California

- Madelyn Garner, Librarian, San Jacinto College, North Campus, Houston, Texas

- Anne Oxenrider, Media Specialist, Dundee High School, Dundee, Michigan

- Charles R. Rodgers, Director of Libraries, Pasco-Hernando Community College, Dade City, Florida

- James N. Zitzelsberger, Library Media Department Chairman, Oshkosh West High School, Oshkosh, Wisconsin

COMMENTS AND SUGGESTIONS

The editors of the *Information Plus Reference Series* welcome your feedback on *Energy: Supplies, Sustainability, and Costs*. Please direct all correspondence to:

Editors
Information Plus Reference Series
27500 Drake Rd.
Farmington Hills, MI 48331-3535

CHAPTER 1
AN ENERGY OVERVIEW

In scientific terms, energy is the capacity for doing work. Humans have learned to harness and use multiple energy sources to get work done. A perfect energy source would be widely available, safe, easy to procure and use, environmentally friendly, provide tremendous amounts of energy, and never run out. As yet, such an energy source has not been found. Instead, there are multiple energy sources with varying levels of suitability. This book will examine the various sources in terms of their availability (supplies), sustainability, and costs.

ENERGY CONVERSIONS

An important law of nature is that energy can neither be created or destroyed, it can only change forms. The most common form of energy conversion facilitated by humans is the conversion of chemical energy (e.g., the energy stored within fuels by virtue of their chemical structures) to thermal energy (i.e., heat). For example, the energy that is stored within wood and other combustible fuels is transferred to heat and light when the fuel is burned. Heat energy can be used directly, such as to heat buildings, or it can be converted to mechanical energy. Imagine a pot of boiling water. The steam rising above the pot is in motion. It is said to have kinetic energy. This energy, when concentrated, can be powerful enough to force movement in machines, such as to turn blades, move pistons, or open valves. In this context, steam is a working fluid. In fact, steam is a popular working fluid because once the desired work has been obtained, the steam can be condensed back to water and reused.

The energy conversion process chemical-to-heat-to-mechanical (C-H-M) is a key element of many energy applications in the modern world including powering vehicles, such as cars, trucks, ships, and bulldozers. The C-H-M process is also the main precursor for producing electricity at power plants. In a very simple power plant, coal or some other fuel is burned to produce steam that turns the blades of a turbine. The turbine rotates a magnet that is nestled within or around coiled wire, generating an electric current in the wire. Thus, mechanical energy is converted to electrical energy. This basic technology has been in use since the 19th century. Hydroelectric power also dates from that century. In this process water is the working fluid that helps convert mechanical energy to electrical energy. During the 20th century scientists developed yet another energy conversion process, one based on nuclear energy (the energy that is held within atomic nuclei). Splitting a uranium atom apart under controlled conditions produces heat that can be used to generate electricity. Wind and solar energy are also increasingly being converted to electrical energy using high-technology devices.

It is important to remember that at each energy conversion stage some of the capacity to do work is lost. For example, some of the heat that is generated by burning fuel escapes to the atmosphere rather than going to heat the desired target. Conversion losses are inevitable. Therefore, humans strive to find energy sources that inherently contain large amounts of energy potential and to minimize energy losses at each conversion step.

ENERGY CONTENT

Energy content is the amount of energy that can be obtained from a given amount of energy source. Heat energy is the key factor by which various energy sources are typically compared. In the United States heat energy is measured using British thermal units (Btu). This term came into use during the 19th century as a benchmark measure: 1 Btu represented the amount of heat that was required to raise the temperature of 1 pound (0.5 kg) of water by 1 degree Fahrenheit (0.6 degrees Celsius). In modern times, the Btu is precisely defined by controlling the measurement variables, for example, the density of the water that is used in the measurement. In countries that rely on the metric system, the joule is the preferred unit for measuring heat, work, and energy; 1 Btu is equivalent to approximately 1,055 joules.

The Btu is a relatively small unit of measure. In "How Can We Compare or Add up Our Energy Consumption?" (March 19, 2012, http://www.eia.gov/energy_in_brief/comparing_energy_consumption.cfm), the Energy Information Administration (EIA) within the U.S. Department of Energy (DOE) explains that 1 Btu is "approximately equal to the amount of energy that comes from burning one wooden kitchen match."

Table 1.1 compares the Btu content of various fossil fuel energy sources and electricity. One ton (0.9 t) of coal provides 19.9 million Btu, compared with only 5.8 million Btu from a barrel (42 gallons [159 L]) of crude oil.

SUSTAINABILITY

In the United States most energy conversion processes begin with fossil fuels—fuels derived from the below-ground remains of prehistoric organisms that became energy enriched over millions of years of being exposed to high temperatures and pressures. Examples include coal, petroleum, and natural gas. These fuels and their derivatives have high energy contents because they contain enormous amounts of carbon that have been stored for millennia. Burning fossil fuels produces large amounts of heat, but also liberates large amounts of carbon into the atmosphere, which is an environmental problem.

Fossil fuels and uranium are called nonrenewable energy sources because there are finite supplies of them. By contrast, renewable energy sources have a theoretically infinite supply. Examples include wind, sunlight, flowing water, and the heat stored beneath the earth's crust. In addition, wood, crops, and combustible wastes produced by humans (e.g., household garbage) are renewable energy sources. Biomass is a term that is used to encompass biologically based materials other than fossil fuels that are used for energy production.

ENERGY PRODUCTION AND CONSUMPTION

People have always found ways to harness energy, such as using animals for transportation and work or inventing machines such as windmills and waterwheels to tap the power of wind and water, respectively. Before the 1800s much of the thermal energy used in the United States for heating and cooking was obtained by burning wood. Lighting was provided by wax candles or by lanterns that burned whale oil or some other kind of animal fat.

The EIA indicates in "A Brief History of Coal Use" (April 24, 2012, http://fossil.energy.gov/education/energylessons/coal/coal_history.html) that commercial coal mining in the United States began in the 1740s. Coal has a much higher heat content than wood, and this energy boost fueled the Industrial Revolution (1760–1848), a period of intense industrialization and modernization. One of the most important innovations of the Industrial Revolution was the steam engine—a machine that converted the thermal energy of burning coal to mechanical energy. Steam engines powered locomotives, ships, and industrial equipment during the 1800s. There were even some steam-powered cars.

The petroleum age began in the United States in 1859, when the first successful modern oil well was drilled. Scientists developed methods to process and refine crude oil into gasoline and other products to power vehicles, ships, and industrial equipment. By the early 20th century petroleum derivatives were in high demand by consumers. Oil also became a commodity that was vital to national security. The United States' large supply of domestic oil was a crucial component to its success in World War II (1939–1945).

Modern U.S. Energy Production

Table 1.2 shows energy production, by fuel, in the United States between 1949 and 2011. The units are quadrillion Btu, with 1 quadrillion Btu equal to 1,000,000,000,000,000 Btu. In 1949 the primary energy sources were coal (12 quadrillion Btu) and crude oil (10.7 quadrillion Btu). Natural gas (5.4 quadrillion Btu) was a distant third. In "The History of Natural Gas" (May 22, 2012, http://www.fossil.energy.gov/education/energylessons/gas/gas_history.html), the EIA explains that natural gas was mostly used during the 1800s as a fuel for street lights. It did not become widely available to individual homes and businesses until after World War II, when extensive pipelines were laid in U.S. cities.

By 1955 crude oil production surpassed that of coal; however, petroleum's reign at the top was short lived. (See Table 1.2.) In 1970 natural gas production surpassed

TABLE 1.1

Btu (British thermal units) content of common energy units

1 barrel (42 gallons) of crude oil	5,800,000 Btu
1 gallon of gasoline	124,238 Btu*
1 gallon of diesel fuel	138,690 Btu
1 gallon of heating oil	138,690 Btu
1 barrel of residual fuel oil	6,287,000 Btu
1 cubic foot of natural gas	1,023 Btu*
1 gallon of propane	91,333 Btu
1 short ton of coal	19,858,000 Btu*
1 kilowatt-hour of electricity	3,412 Btu

*Based on U.S. consumption, 2011
Note: The Btu content of each fuel reflects the average energy content for fuels consumed in the United States.

SOURCE: Adapted from "Btu Content of Common Energy Units," in *Energy Units and Calculators Explained*, U.S. Energy Information Administration, September 24, 2012, http://www.eia.gov/energyexplained/index.cfm?page=about_energy_units (accessed October 4, 2012)

TABLE 1.2

Primary energy production, by source, selected years, 1949–2011

[Quadrillion Btu]

Year	Coal[b]	Natural gas (dry)	Crude oil[c]	NGPL[d]	Total (Fossil fuels)	Nuclear electric power[e]	Hydroelectric power[f]	Geothermal[g]	Solar/PV[h]	Wind[i]	Biomass[j]	Total (Renewable energy[a])	Total
1949	11.974	5.377	10.683	0.714	28.748	0.000	1.425	NA	NA	NA	1.549	2.974	31.722
1950	14.060	6.233	11.447	0.823	32.563	0.000	1.415	NA	NA	NA	1.562	2.978	35.540
1955	12.370	9.345	14.410	1.240	37.364	0.000	1.360	NA	NA	NA	1.424	2.784	40.148
1960	10.817	12.656	14.935	1.461	39.869	0.006	1.608	(s)	NA	NA	1.320	2.928	42.803
1965	13.055	15.775	16.521	1.883	47.235	0.043	2.059	0.002	NA	NA	1.335	3.396	50.674
1970	14.607	21.666	20.401	2.512	59.186	0.239	2.634	0.006	NA	NA	1.431	4.070	63.495
1975	14.989	19.640	17.729	2.374	54.733	1.900	3.155	0.034	NA	NA	1.499	4.687	61.320
1976	15.654	19.480	17.262	2.327	54.723	2.111	2.976	0.038	NA	NA	1.713	4.727	61.561
1977	15.755	19.565	17.454	2.327	55.101	2.702	2.333	0.037	NA	NA	1.838	4.209	62.012
1978	14.910	19.485	18.434	2.245	55.074	3.024	2.937	0.031	NA	NA	2.038	5.005	63.104
1979	17.540	20.076	18.104	2.286	58.006	2.776	2.931	0.040	NA	NA	2.152	5.123	65.904
1980	18.598	19.908	18.249	2.254	59.008	2.739	2.900	0.053	NA	NA	2.476	5.428	67.175
1981	18.377	19.699	18.146	2.307	58.529	3.008	2.758	0.059	NA	NA	2.596	5.414	66.951
1982	18.639	18.319	18.309	2.191	57.458	3.131	3.266	0.051	NA	NA	2.663	5.980	66.569
1983	17.247	16.593	18.392	2.184	54.416	3.203	3.527	0.064	NA	NA	2.904	6.496	64.114
1984	19.719	18.008	18.848	2.274	58.849	3.553	3.386	0.081	NA	NA	2.971	6.438	68.840
1985	19.325	16.980	18.992	2.241	57.539	4.076	2.970	0.097	(s)	(s)	3.016	6.084	67.698
1986	19.509	16.541	18.376	2.149	56.575	4.380	3.071	0.108	(s)	(s)	2.932	6.111	67.066
1987	20.141	17.136	17.675	2.215	57.167	4.754	2.635	0.112	(s)	(s)	2.875	5.622	67.542
1988	20.738	17.599	17.279	2.260	57.875	5.587	2.334	0.106	(s)	(s)	3.016	5.457	68.919
1989	R21.360	17.847	16.117	2.158	57.483	5.602	2.837	0.162	(s)	0.022	3.159	6.235	69.320
1990	22.488	18.326	15.571	2.175	58.560	6.104	3.046	0.171	0.055	0.029	2.735	6.041	70.705
1991	21.636	18.229	15.701	2.306	57.872	6.422	3.016	0.178	0.059	0.031	2.782	6.069	70.362
1992	21.694	18.375	15.223	2.363	57.655	6.479	2.617	0.179	0.062	0.030	2.932	5.821	R69.956
1993	20.336	18.584	14.494	2.408	55.822	6.410	2.892	0.186	0.064	0.031	2.908	6.083	68.315
1994	22.202	19.348	14.103	2.391	58.044	6.694	2.683	0.173	0.066	0.036	3.028	5.988	70.726
1995	22.130	19.082	13.887	2.442	57.540	7.075	3.205	0.152	0.069	0.033	3.099	6.558	71.174
1996	22.790	19.344	13.723	2.530	58.387	7.087	3.590	0.163	0.070	0.033	3.155	7.012	72.486
1997	23.310	19.394	13.658	2.495	58.857	6.597	3.640	0.167	0.070	0.034	3.108	7.018	72.472
1998	24.045	19.613	13.235	2.420	59.314	7.068	3.297	0.168	0.069	0.031	2.929	6.494	72.876
1999	23.295	19.341	12.451	2.528	57.614	7.610	3.268	0.171	0.068	0.046	2.965	6.517	71.742
2000	22.735	19.662	12.358	2.611	57.366	7.862	2.811	0.164	R0.066	0.057	3.006	6.104	71.332
2001	U23.547	20.166	12.282	2.547	58.541	8.029	2.242	0.164	0.064	0.070	2.624	5.164	71.735
2002	22.732	R19.382	12.163	2.559	R56.837	8.145	2.689	0.171	0.063	0.105	2.705	5.734	R70.716
2003	22.094	19.633	12.026	2.346	56.099	7.959	2.825	0.175	0.062	0.115	2.805	5.982	70.040
2004	22.852	19.074	11.503	2.466	55.895	8.222	2.690	0.178	0.063	0.142	2.998	6.070	70.188
2005	23.185	18.556	10.963	2.334	55.038	8.161	2.703	0.181	0.063	0.178	3.104	6.229	R69.428
2006	23.790	19.022	10.801	2.356	55.968	8.215	2.869	0.181	0.068	0.264	R3.216	R6.599	R70.782
2007	23.493	R19.786	10.721	2.409	R56.409	8.455	2.446	0.186	0.076	0.341	R3.461	R6.509	R71.373
2008	23.851	20.703	10.509	2.419	57.482	8.427	2.511	0.192	0.089	0.546	R3.864	R7.202	R73.111
2009	R21.624	R21.139	11.348	2.574	R56.685	8.356	2.669	0.200	0.098	0.721	R3.928	R7.616	R72.657
2010	R22.038	R21.823	R11.593	R2.781	R58.235	R8.434	2.539	R0.208	R0.126	R0.923	R4.341	R8.136	R74.806
2011[P]	22.181	23.506	11.986	2.928	60.601	8.259	3.171	0.226	0.158	1.168	4.511	9.236	78.096

TABLE 1.2

Primary energy production, by source, selected years, 1949–2011 [CONTINUED]

[Quadrillion Btu]

[a]Most data are estimates.
[b]Beginning in 1989, includes waste coal supplied. Beginning in 2001, also includes a small amount of refuse recovery.
[c]Includes lease condensate.
[d]Natural gas plant liquids.
[e]Nuclear electricity net generation (converted to Btu using the nuclear heat rate).
[f]Conventional hydroelectricity net generation (converted to Btu using the fossil-fuels heat rate).
[g]Geothermal electricity net generation (converted to Btu using the fossil-fuels heat rate), and geothermal heat pump and direct use energy.
[h]Solar thermal and photovoltaic (PV) electricity net generation (converted to Btu using the fossil-fuels heat rate), and solar thermal direct use energy.
[i]Wind electricity net generation (converted to Btu using the fossil-fuels heat rate).
[j]Wood and wood-derived fuels, biomass waste, and total biomass inputs to the production of fuel ethanol and biodiesel.
R = Revised. P = Preliminary. NA = Not available. (s) = Less than 0.0005 quadrillion Btu.
Btu = British thermal unit.
Note: Totals may not equal sum of components due to independent rounding.

SOURCE: "Table 1.2. Primary Energy Production by Source, Selected Years, 1949–2011 (Quadrillion Btu)," in *Annual Energy Review 2011*, U.S. Energy Information Administration, September 27, 2012. http://www.eia.gov/totalenergy/data/annual/pdf/aer.pdf (accessed October 3, 2012)

4   An Energy Overview

Energy

that of crude oil. During the early 1980s these three fossil fuels were roughly equal in terms of domestic production. Over the following decades domestic production fell dramatically for crude oil, but grew stronger for coal and natural gas.

Figure 1.1 illustrates that, for decades, fossil fuel production vastly overshadowed the production of other energy sources. Among the latter, the top competitors were nuclear electric power and renewable energy, specifically biomass and hydroelectric power:

- Nuclear electric power—production increased dramatically during the 1970s, from 0.2 quadrillion Btu in 1970 to 3 quadrillion Btu in 1978. (See Table 1.2.) It continued to rise, doubling to 6.1 quadrillion Btu by 1990 and reaching its highest production level of 8.5 quadrillion Btu in 2007. Production declined over subsequent years, dropping to 8.3 quadrillion Btu in 2011.

- Biomass—production has risen and fallen over the decades. It rose above 3 quadrillion Btu during the late 1980s, mid-1990s, and 2000, only to decline in subsequent years. Between 2001 and 2011 biomass production experienced a sustained production increase, reaching a record 4.5 quadrillion Btu in 2011.

- Hydroelectric power—production rose to its highest level during the mid-1990s, peaking at 3.6 quadrillion Btu in 1997, and then declined to less than 3 quadrillion Btu per year for more than a decade. Hydroelectric power production rebounded somewhat in 2011, to 3.2 quadrillion Btu.

FIGURE 1.1

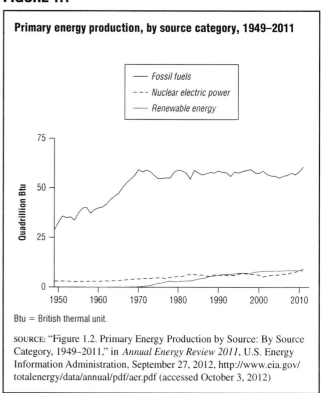

Primary energy production, by source category, 1949–2011

Btu = British thermal unit.

SOURCE: "Figure 1.2. Primary Energy Production by Source: By Source Category, 1949–2011," in *Annual Energy Review 2011*, U.S. Energy Information Administration, September 27, 2012, http://www.eia.gov/totalenergy/data/annual/pdf/aer.pdf (accessed October 3, 2012)

Figure 1.2 shows primary energy production, by source, for 2011. At 24 quadrillion Btu, natural gas accounted for more energy production in the United States during 2011 than any other energy source. Energy produced from coal was a close second (22 quadrillion

FIGURE 1.2

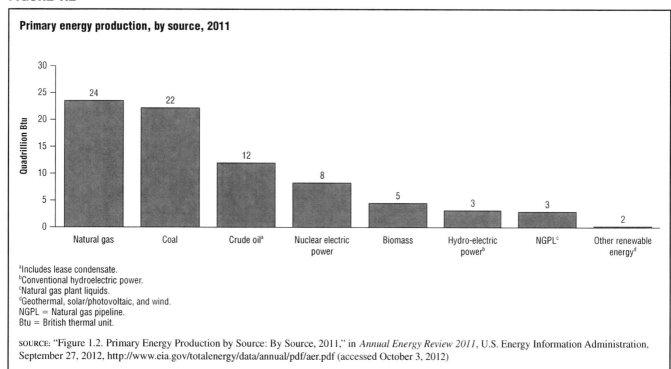

Primary energy production, by source, 2011

aIncludes lease condensate.
bConventional hydroelectric power.
cNatural gas plant liquids.
dGeothermal, solar/photovoltaic, and wind.
NGPL = Natural gas pipeline.
Btu = British thermal unit.

SOURCE: "Figure 1.2. Primary Energy Production by Source: By Source, 2011," in *Annual Energy Review 2011*, U.S. Energy Information Administration, September 27, 2012, http://www.eia.gov/totalenergy/data/annual/pdf/aer.pdf (accessed October 3, 2012)

Btu), followed by crude oil (12 quadrillion Btu) and nuclear electric power (8 quadrillion Btu). The total domestic energy production of the United States (the amount of fossil fuels and other forms of energy that were mined, pumped, or otherwise originated in the United States) more than doubled, from 31.7 quadrillion Btu in 1949 to 78.1 quadrillion Btu in 2011. (See Table 1.2.)

Modern U.S. Energy Consumption

Even though the total domestic energy production more than doubled between 1949 and 2011, the total domestic energy consumption (the amount of energy used in the United States) more than tripled during that period, from 32 quadrillion Btu in 1949 to 97.3 quadrillion Btu in 2011. (See Table 1.3 and Figure 1.3.) This increase did not, however, happen in an even progression. Energy consumption doubled between 1949 and 1970 to 67.8 quadrillion Btu. It wavered up and down during the 1970s and then began to climb again, peaking at 101.3 quadrillion Btu in 2007. That same year witnessed the beginning of the Great Recession, which lasted from December 2007 to June 2009 and featured reduced consumer demand for many products, including energy products. After falling dramatically in 2008, energy consumption began to rise again in 2009 and was at 97.3 quadrillion Btu in 2011.

One of the reasons that energy consumption has increased in the United States is a growing population. According to the U.S. Census Bureau, in *Measuring America: The Decennial Censuses from 1790 to 2000* (September 2002, http://www.census.gov/prod/2002pubs/po l02marv.pdf), the U.S. population numbered 151.3 million in 1950. In "Texas Gains the Most in Population since the Census" (December 21, 2011, http://www.census .gov/newsroom/releases/archives/population/cb11-215 .html), the Census Bureau estimates the U.S. population was 311.6 million in mid-2011. Thus, the population more than doubled between 1950 and 2011. Energy consumption grew even faster, nearly tripling during the same period, from 34.6 quadrillion Btu in 1950 to 97.3 quadrillion Btu in 2011. (See Table 1.3.)

The EIA divides in *Annual Energy Review 2011* (September 2012, http://www.eia.gov/totalenergy/data/annual/ pdf/aer.pdf) U.S. energy consumers into five broad sectors: electric power, residential, industrial, commercial, and transportation.

ELECTRIC POWER SECTOR. This sector consumes energy at facilities it operates to sell electricity to the public. Examples include electric utilities and combined heat-and-power plants that sell heat and electricity. It should be noted that this sector does not include power plants that are operated by factories and other industrial enterprises that produce electricity or electricity and heat for their own use. Figure 1.4 shows the primary fuels that were used by the electric power sector between 1949 and

2011. Coal has historically been the main fuel source for electric power generation.

The EIA considers the electric power sector an intermediary in the energy supply chain, not an end user of energy. As will be explained in Chapter 8, electricity generation is inherently inefficient; energy content is lost as fuels are converted to electricity and as the electricity travels across transmission lines and distribution systems. Thus, the heat content of the energy sources that enter the electric power sector is far greater than the heat content of the electricity that is delivered to end users. The difference is called electrical losses. The EIA tracks electrical losses when it calculates energy consumption by end-use sectors.

RESIDENTIAL SECTOR. This end-use sector consumes energy in residential living quarters. Common uses include space heating, water heating, air conditioning, lighting, refrigeration, cooking, and running other appliances. Figure 1.5 shows the major energy sources that were consumed by this sector between 1949 and 2011. Natural gas has historically been the most-used source; however, its dominance was matched by electricity during the first decade of the 21st century. Petroleum, renewable energy, and coal are minor energy sources to this sector.

INDUSTRIAL SECTOR. Energy consumption by this end-use sector powers equipment and facilities that are engaged in producing, processing, or assembling goods. Covered industries include manufacturing, mining (including oil and gas extraction), and construction activities. Agriculture, forestry, fishing, and hunting enterprises are also included. Major energy uses within the industrial sector are for process heat and for cooling and powering machinery. In addition, the EIA notes that fossil fuels may be used as raw material inputs to manufactured products. This sector includes generators that produce electricity and/or useful thermal output primarily to support industrial activities. Figure 1.6 shows the major energy sources that were consumed by the industrial sector between 1949 and 2011. During the 1950s petroleum and natural gas replaced coal as the preferred fuel. In 2011 petroleum and natural gas were the top fuels and were used roughly equally. Electricity, renewable energy, and coal played much smaller roles.

COMMERCIAL SECTOR. This end-use sector includes the equipment and facilities that are operated by businesses; federal, state, and local governments; and other private and public organizations, such as religious, social, or fraternal groups. The commercial sector includes institutional living quarters and sewage treatment facilities. Common energy uses include space heating, water heating, air conditioning, lighting, refrigeration, cooking, and running equipment. The EIA notes that this sector includes generators that produce electricity and/or useful

TABLE 1.3

Primary energy consumption, by source, selected years, 1949–2011

[Quadrillion Btu]

Year	Fossil fuels					Nuclear electric power	Renewable Energy[a]					Electricity net imports[c]	Total
							Noncombustible[b]						
	Coal	Coal coke net imports[c]	Natural gas[d]	Petroleum[e]	Total		Captured energy	Adjustment for fossil fuel equivalence	Total	Biomass	Total		
1949	11.981	−0.007	5.145	11.883	29.002	0.000	0.323	1.101	1.425	1.549	2.974	0.005	31.982
1950	12.347	0.001	5.968	13.315	31.632	0.000	0.344	1.071	1.415	1.562	2.978	0.006	34.616
1955	11.167	−0.010	8.998	17.255	37.410	0.000	0.397	0.963	1.360	1.424	2.784	0.014	40.208
1960	9.838	−0.006	12.385	19.919	42.137	0.006	0.510	1.098	1.608	1.320	2.928	0.015	45.086
1965	11.581	−0.018	15.769	23.246	50.577	0.043	0.673	1.388	2.061	1.335	3.396	(s)	54.015
1970	12.265	−0.058	21.795	29.521	63.522	0.239	0.858	1.781	2.639	1.431	4.070	0.007	67.838
1975	12.663	0.014	19.948	32.732	65.357	1.900	1.045	1.781	3.188	1.499	4.687	0.021	71.965
1976	13.584	(s)	20.345	35.178	69.107	2.111	0.991	2.022	3.014	1.713	4.727	0.029	75.975
1977	13.922	0.015	19.931	37.124	70.991	2.702	0.775	1.595	2.371	1.838	4.209	0.059	77.961
1978	13.766	0.125	20.000	37.963	71.854	3.024	0.977	1.990	2.968	2.038	5.005	0.067	79.950
1979	15.040	0.063	20.666	37.122	72.891	2.776	0.979	1.992	2.971	2.152	5.123	0.069	80.859
1980	15.423	−0.035	20.235	34.205	69.828	2.739	0.970	1.983	2.953	2.476	5.428	0.071	78.067
1981	15.908	−0.016	19.747	31.932	67.571	3.008	0.920	1.898	2.817	2.596	5.414	0.113	76.106
1982	15.322	−0.022	18.356	30.232	63.888	3.131	1.082	2.234	3.316	2.663	5.980	0.100	73.099
1983	15.894	−0.016	17.221	30.052	63.152	3.203	1.165	2.426	3.591	2.904	6.496	0.121	72.971
1984	17.071	−0.011	18.394	31.053	66.506	3.553	1.133	2.334	3.467	2.971	6.438	0.135	76.632
1985	17.478	−0.013	17.703	30.925	66.093	4.076	1.002	2.066	3.068	3.016	6.084	0.140	76.392
1986	17.260	−0.017	16.591	32.198	66.033	4.380	1.038	2.141	3.179	2.932	6.111	0.122	76.647
1987	18.008	0.009	17.640	32.864	68.521	4.754	0.900	1.847	2.747	2.875	5.622	0.158	79.054
1988	18.846	0.040	18.448	34.223	71.557	5.587	0.807	1.634	2.441	3.016	5.457	0.108	82.709
1989	19.070	0.030	19.602	34.209	72.911	5.602	1.048	2.028	3.076	3.159	6.235	0.037	84.786
1990	19.173	0.005	19.603	33.552	72.332	6.104	1.128	2.177	3.306	2.735	6.041	0.008	84.485
1991	18.992	0.010	20.033	32.846	71.880	6.422	1.121	2.166	3.287	2.782	6.069	0.067	84.438
1992	19.122	0.035	20.714	33.525	73.396	6.479	1.001	1.889	2.890	2.932	5.821	0.087	85.783
1993	19.835	0.027	21.229	33.745	74.836	6.410	1.100	2.074	3.174	2.908	6.083	0.095	87.424
1994	19.909	0.058	21.728	34.561	76.256	6.694	1.030	1.930	2.961	3.028	5.988	0.153	89.091
1995	20.089	0.061	22.671	34.438	77.259	7.075	1.197	2.262	3.459	3.101	6.560	0.134	91.029
1996	21.002	0.023	23.085	35.675	79.785	7.087	1.326	2.530	3.857	3.157	7.014	0.137	94.022
1997	21.445	0.046	23.223	36.159	80.873	6.597	1.360	2.550	3.910	3.105	7.016	0.116	94.602
1998	21.656	0.067	22.830	36.816	81.369	7.068	1.247	2.318	3.565	2.927	6.493	0.088	95.018
1999	21.623	0.058	22.909	37.838	82.427	7.610	1.240	2.312	3.552	2.963	6.516	0.099	96.652
2000	22.580	0.065	23.824	38.262	84.731	7.862	1.090	2.008	3.098	3.008	6.106	0.115	98.814
2001	21.914	0.029	22.773	38.186	82.902	8.029	0.893	1.647	2.540	2.622	5.163	0.075	96.168
2002	21.904	0.061	R23.510	38.224	R83.699	8.145	1.070	1.959	3.029	2.701	5.729	0.072	R97.645
2003	22.321	0.051	22.831	38.811	84.014	7.959	1.114	2.062	3.176	2.807	5.983	0.022	97.978
2004	22.466	0.138	R22.923	40.292	R85.819	8.222	1.103	1.969	3.073	3.010	6.082	0.039	R100.162
2005	22.797	0.044	R22.565	40.388	R85.794	8.161	1.127	1.998	3.125	R3.117	6.242	0.085	R100.282
2006	22.447	0.061	R22.239	39.955	R84.702	8.215	1.229	2.153	3.382	R3.267	R6.649	0.063	99.629
2007	22.749	0.025	R23.663	39.774	R86.211	8.455	1.125	1.924	3.048	R3.474	R6.523	0.107	R101.296
2008	22.385	0.041	R23.843	37.280	R83.549	8.427	1.238	2.099	3.338	R3.849	R7.186	0.112	99.275
2009	19.692	−0.024	23.416	35.403	78.488	8.356	1.382	2.306	3.688	R3.912	R7.600	0.116	94.559
2010	R20.850	−0.006	R24.256	R36.010	R81.109	R8.434	R1.440	R2.355	R3.796	4.294	R8.090	R0.089	97.722
2011P	19.643	0.011	24.843	35.283	79.779	8.259	1.785	2.939	4.724	4.411	9.135	0.127	97.301

TABLE 1.3

Primary energy consumption, by source, selected years, 1949–2011 [CONTINUED]

[Quadrillion Btu]

[a]Most data are estimates.
[b]Conventional hydroelectric power, geothermal, solar thermal, photovoltaic, and wind.
[c]Net imports equal imports minus exports. A minus sign indicates exports are greater than imports.
[d]Natural gas only; excludes supplemental gaseous fuels.
[e]Petroleum products supplied, including natural gas plant liquids and crude oil burned as fuel. Does not include biofuels that have been blended with petroleum—biofuels are included in "Biomass." For petroleum, product supplied is used as an approximation of petroleum consumption.
R = Revised. P = Preliminary. (s) = Less than 0.0005 and greater than −0.0005 quadrillion Btu.
Btu = British thermal unit.
Note: Totals may not equal sum of components due to independent rounding.

SOURCE: "Table 1.3. Primary Energy Consumption Estimates by Source, Selected Years, 1949–2011 (Quadrillion Btu)," in *Annual Energy Review 2011*, U.S. Energy Information Administration, September 27, 2012. http://www.eia.gov/totalenergy/data/annual/pdf/aer.pdf (accessed October 3, 2012)

FIGURE 1.3

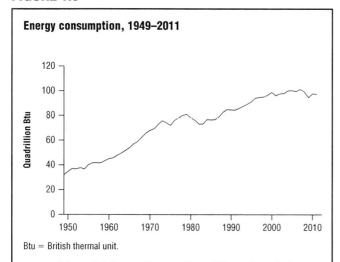

Energy consumption, 1949–2011

Btu = British thermal unit.

SOURCE: "Figure 1.5. Energy Consumption and Expenditures Indicators Estimates: Energy Consumption, 1949–2011 (Quadrillion Btu)," in *Annual Energy Review 2011*, U.S. Energy Information Administration, September 27, 2012, http://www.eia.gov/totalenergy/data/annual/pdf/ aer.pdf (accessed October 3, 2012)

FIGURE 1.4

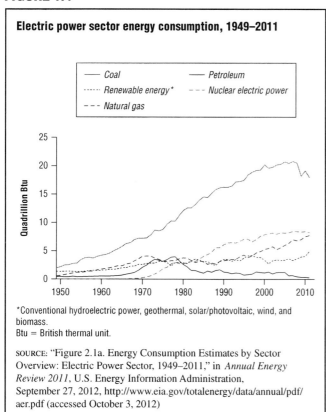

Electric power sector energy consumption, 1949–2011

*Conventional hydroelectric power, geothermal, solar/photovoltaic, wind, and biomass.
Btu = British thermal unit.

SOURCE: "Figure 2.1a. Energy Consumption Estimates by Sector Overview: Electric Power Sector, 1949–2011," in *Annual Energy Review 2011*, U.S. Energy Information Administration, September 27, 2012, http://www.eia.gov/totalenergy/data/annual/pdf/ aer.pdf (accessed October 3, 2012)

FIGURE 1.5

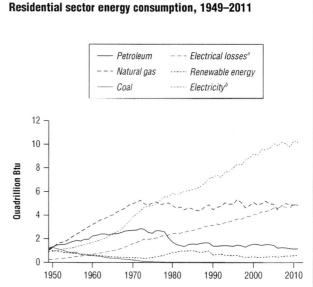

Residential sector energy consumption, 1949–2011

aElectrical system energy losses associated with the generation, transmission, and distribution of energy in the form of electricity.
bElectricity retail sales.
Btu = British thermal unit.

SOURCE: "Figure 2.1b. Energy Consumption Estimates by End-Use Sector, 1949–2011: Residential, by Major Source," in *Annual Energy Review 2011*, U.S. Energy Information Administration, September 27, 2012, http://www.eia.gov/totalenergy/data/annual/pdf/aer.pdf (accessed October 3, 2012)

FIGURE 1.6

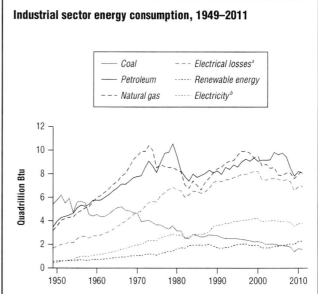

Industrial sector energy consumption, 1949–2011

aElectrical system energy losses associated with the generation, transmission, and distribution of energy in the form of electricity.
bElectricity retail sales.
Btu = British thermal unit.

SOURCE: "Figure 2.1b. Energy Consumption Estimates by End-Use Sector, 1949–2011: Industrial, by Major Source," in *Annual Energy Review 2011*, U.S. Energy Information Administration, September 27, 2012, http://www.eia.gov/totalenergy/data/annual/pdf/aer.pdf (accessed October 3, 2012)

thermal output primarily to support the activities of commercial establishments. Figure 1.7 shows that electricity and natural gas have been the preferred fuels in this sector for decades. In 2011 electricity was the main energy source, followed by natural gas, petroleum, renewable energy, and coal.

FIGURE 1.7

Commercial sector energy consumption, 1949–2011

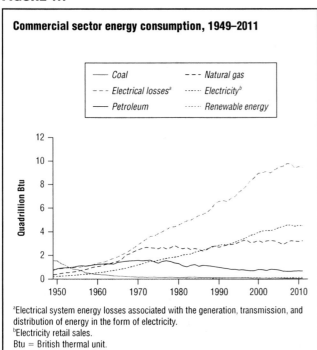

aElectrical system energy losses associated with the generation, transmission, and distribution of energy in the form of electricity.
bElectricity retail sales.
Btu = British thermal unit.

SOURCE: "Figure 2.1b. Energy Consumption Estimates by End-Use Sector, 1949–2011: Commercial, by Major Source," in *Annual Energy Review 2011*, U.S. Energy Information Administration, September 27, 2012, http://www.eia.gov/totalenergy/data/annual/pdf/aer.pdf (accessed October 3, 2012)

FIGURE 1.8

Transportation sector energy consumption, 1949–2011

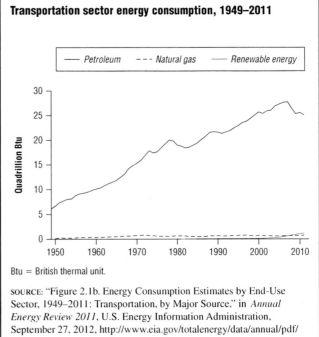

Btu = British thermal unit.

SOURCE: "Figure 2.1b. Energy Consumption Estimates by End-Use Sector, 1949–2011: Transportation, by Major Source," in *Annual Energy Review 2011*, U.S. Energy Information Administration, September 27, 2012, http://www.eia.gov/totalenergy/data/annual/pdf/aer.pdf (accessed October 3, 2012)

FIGURE 1.9

Primary energy consumption, by major source, 1949–2011

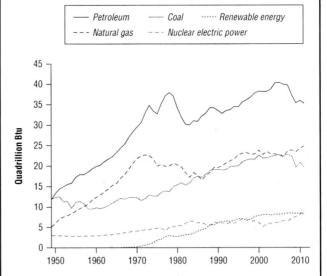

*Petroleum products supplied, including natural gas plant liquids and crude oil burned as fuel. Does not include biofuels that have been blended with petroleum—biofuels are included in "Renewable Energy." For petroleum, product supplied is used as an approximation of petroleum consumption.
Btu = British thermal unit.

SOURCE: "Figure 1.3. Primary Energy Consumption Estimates by Source: By Major Source, 1949–2011," in *Annual Energy Review 2011*, U.S. Energy Information Administration, September 27, 2012, http://www.eia.gov/totalenergy/data/annual/pdf/aer.pdf (accessed October 3, 2012)

TRANSPORTATION SECTOR. Energy consumed by this end-use sector is for vehicles whose primary purpose is transporting people and/or goods from place to place. These vehicles include automobiles; trucks; buses; motorcycles; trains, subways, and other rail vehicles; aircraft; and ships, barges, and other waterborne vehicles. It should be noted that vehicles whose primary purpose is not transportation (e.g., bulldozers, tractors, and forklifts) are classified in the sector of their primary use. Petroleum dominates, by far, the energy consumption in this sector. (See Figure 1.8.) Other energy sources are minor players.

SECTORS AND ENERGY SOURCES. The EIA explains that total energy consumption is determined by summing primary energy consumption by the end-use sectors, electricity retail sales, and electrical losses. Figure 1.9 shows primary energy consumption by energy source between 1949 and 2011. Petroleum dominated over the other energy sources by a large margin throughout most of the period. Natural gas and coal have steadfastly maintained second and third place, respectively, in consumption. Nuclear electric power and renewable energy have played smaller roles in the nation's overall energy consumption. Figure 1.10 lays out the relationships between the energy sources and the energy-using sectors in 2011.

Historically, industry has been the largest energy consumer of the end-use sectors. (See Figure 1.11.) In 2011 industry used 30.6 quadrillion Btu, compared with

FIGURE 1.10

Primary energy consumption, by source and by sector, 2011

[Quadrillion Btu]

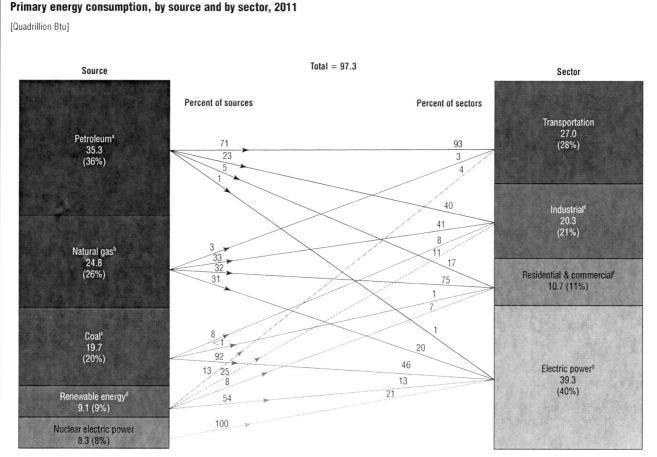

Total = 97.3

Source

Sector

Percent of sources

Percent of sectors

Petroleum[a]
35.3
(36%)

Natural gas[b]
24.8
(26%)

Coal[c]
19.7
(20%)

Renewable energy[d]
9.1 (9%)

Nuclear electric power
8.3 (8%)

Transportation
27.0
(28%)

Industrial[e]
20.3
(21%)

Residential & commercial[f]
10.7 (11%)

Electric power[g]
39.3
(40%)

[a]Does not include biofuels that have been blended with petroleum—biofuels are included in "Renewable energy."
[b]Excludes supplemental gaseous fuels.
[c]Includes less than 0.1 quadrillion Btu of coal coke net imports.
[d]Conventional hydroelectric power, geothermal, solar/photovoltaic, wind, and biomass.
[e]Includes industrial combined-heat-and-power (CHP) and industrial electricity-only plants.
[f]Includes commercial combined-heat-and-power (CHP) and commercial electricity-only plants.
[g]Electricity-only and combined-heat-and-power (CHP) plants whose primary business is to sell electricity, or electricity and heat, to the public. Includes 0.1 quadrillion Btu of electricity net imports not shown under "Source."
Notes: Primary energy in the form that it is first accounted for in a statistical energy balance, before any transformation to secondary or tertiary forms of energy (for example, coal is used to generate electricity). Sum of components may not equal total due to independent rounding.
Btu = British thermal unit.

SOURCE: "Figure 2.0. Primary Energy Consumption by Source and Sector, 2011 (Quadrillion Btu)," in *Annual Energy Review 2011*, U.S. Energy Information Administration, September 27, 2012, http://www.eia.gov/totalenergy/data/annual/pdf/aer.pdf (accessed October 3, 2012)

27.1 quadrillion Btu for the transportation sector, 21.6 quadrillion Btu for the residential sector, and 18 quadrillion Btu for the commercial sector.

Figure 1.12 shows primary and total energy consumption in 2011 for all energy-consuming sectors (i.e., the end-use sectors plus the electric power sector). This chart illustrates that the differences between primary and total energy consumption for each end-use sector sum to the energy consumed by the electric power sector.

ENERGY SELF-SUFFICIENCY

Before the 20th century, the United States was largely energy self-sufficient, in that it produced enough energy domestically to satisfy domestic demand. Through the

1950s domestic energy production and consumption were nearly equal. (See Figure 1.13.) During the 1960s consumption slightly outpaced production. During the 1970s the gap widened considerably, narrowed somewhat during the early 1980s, and thereafter widened year after year. By the beginning of the 21st century the gap between domestic energy production and consumption was quite significant.

Since the 1970s energy imports (particularly of petroleum) have been used to close the gap between domestic energy production and consumption. Petroleum imports have historically encompassed the vast majority of total energy imports. (See Figure 1.14.) The United States' dependence on other countries for oil has created economic and political problems that began in the 1970s. During that decade the nation experienced an energy

FIGURE 1.11

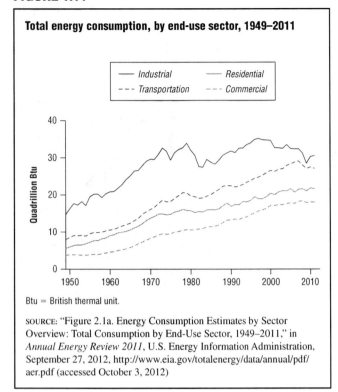

Total energy consumption, by end-use sector, 1949–2011

— Industrial   — Residential
- - - Transportation   - - - Commercial

Btu = British thermal unit.

SOURCE: "Figure 2.1a. Energy Consumption Estimates by Sector Overview: Total Consumption by End-Use Sector, 1949–2011," in *Annual Energy Review 2011*, U.S. Energy Information Administration, September 27, 2012, http://www.eia.gov/totalenergy/data/annual/pdf/aer.pdf (accessed October 3, 2012)

crisis because it was hugely dependent on foreign oil, particularly from countries in the Middle East. A political disagreement led some of those countries to temporarily limit the amount of oil that was sold to the United States.

Lower supply in the face of growing demand sent prices soaring for gasoline and other oil products. The energy embargo had severe repercussions on the overall economy. Energy self-sufficiency became a new national goal, but, as shown in Figure 1.13, this goal has been difficult to achieve.

The United States also exports some energy commodities. U.S. petroleum exports totaled 5.9 quadrillion Btu in 2011. (See Figure 1.15.) According to the EIA, in *Annual Energy Review 2011*, U.S. petroleum exports are almost all petroleum products, such as unfinished oils, pentane, and gasoline blending components. The United States also exported 2.8 quadrillion Btu of coal in 2011.

Figure 1.16 shows domestic energy production, import, export, and consumption flows in 2011. Overall, domestic production was 78.1 quadrillion Btu. Imports added another 28.6 quadrillion Btu, mostly from crude oil and petroleum products. Thus, the total U.S. energy supply was 107.7 quadrillion Btu. Approximately 10.4 quadrillion Btu of the supply was exported. U.S. energy exports in 2011 consisted of 5.9 quadrillion Btu of petroleum products and 4.5 quadrillion Btu of other commodities, such as natural gas, coal, coal coke, biofuels, and electricity.

Figure 1.17 gives the overall picture for imports and exports in the nation's energy flows in 2011. Overall, 29 quadrillion Btu of energy was imported, which represented 27% of the total energy supplied that year. By

FIGURE 1.12

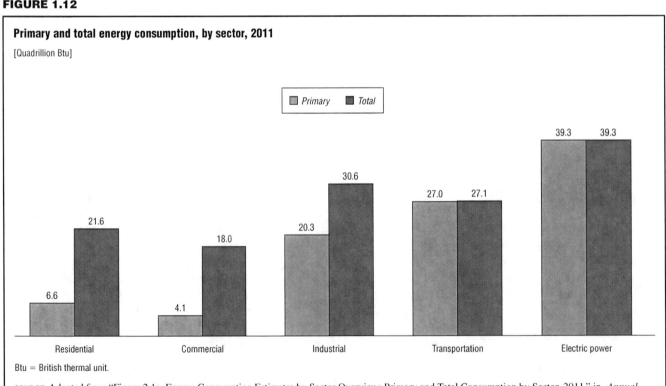

Primary and total energy consumption, by sector, 2011

[Quadrillion Btu]

Primary   Total

	Primary	Total
Residential	6.6	21.6
Commercial	4.1	18.0
Industrial	20.3	30.6
Transportation	27.0	27.1
Electric power	39.3	39.3

Btu = British thermal unit.

SOURCE: Adapted from "Figure 2.1a. Energy Consumption Estimates by Sector Overview: Primary and Total Consumption by Sector, 2011," in *Annual Energy Review 2011*, U.S. Energy Information Administration, September 27, 2012, http://www.eia.gov/totalenergy/data/annual/pdf/aer.pdf (accessed October 3, 2012)

FIGURE 1.13

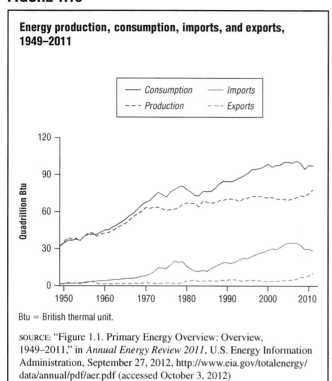

Energy production, consumption, imports, and exports, 1949–2011

Btu = British thermal unit.

SOURCE: "Figure 1.1. Primary Energy Overview: Overview, 1949–2011," in *Annual Energy Review 2011*, U.S. Energy Information Administration, September 27, 2012, http://www.eia.gov/totalenergy/data/annual/pdf/aer.pdf (accessed October 3, 2012)

FIGURE 1.14

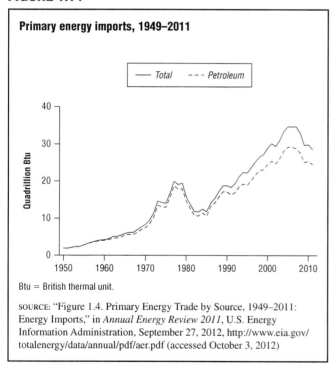

Primary energy imports, 1949–2011

Btu = British thermal unit.

SOURCE: "Figure 1.4. Primary Energy Trade by Source, 1949–2011: Energy Imports," in *Annual Energy Review 2011*, U.S. Energy Information Administration, September 27, 2012, http://www.eia.gov/totalenergy/data/annual/pdf/aer.pdf (accessed October 3, 2012)

FIGURE 1.15

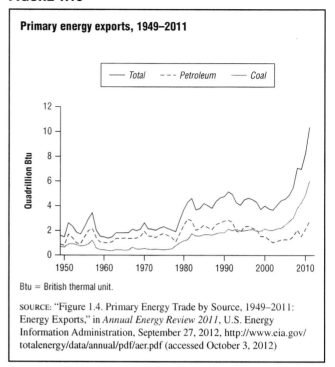

Primary energy exports, 1949–2011

Btu = British thermal unit.

SOURCE: "Figure 1.4. Primary Energy Trade by Source, 1949–2011: Energy Exports," in *Annual Energy Review 2011*, U.S. Energy Information Administration, September 27, 2012, http://www.eia.gov/totalenergy/data/annual/pdf/aer.pdf (accessed October 3, 2012)

contrast, exports of 10 quadrillion Btu accounted for 9% of total energy consumption.

ENERGY COSTS

Energy costs, like the costs for all commodities, depend on supply and demand factors. These factors are quite complex, and some of the major drivers are discussed in this chapter. In addition, source-specific supply and demand factors are discussed in detail in Chapter 2 for oil, in Chapter 3 for natural gas, in Chapter 4 for coal, in Chapter 5 for nuclear energy, in Chapter 6 for renewable energy, and in Chapter 8 for electricity.

Energy costs can be divided into two broad categories: market costs and external costs.

Market Costs

Market costs are the prices paid in the marketplace for energy products. The EIA tracks in *Annual Energy Review 2011* various kinds of market costs, including energy production costs, import and export energy costs, and consumer costs for energy. Source-specific costs are discussed in detail in the relevant chapters. However, some general information about costing trends is presented here.

The EIA notes that the nation's overall price tag for total energy purchased by consumers increased modestly from less than $2 per million Btu in 1970 to around $8 per million Btu during the early 1980s. It stayed at this level well into the 1990s before it began to creep upward. The first decade of the 21st century saw an unprecedented rise in energy costs. In 2008 the price peaked above $21 per million Btu. It then declined slightly, dropping to around $19 per million Btu in 2010. It should be noted that these are nominal prices, meaning that they do not reflect the effects of inflation over time. According to the

FIGURE 1.16

Energy flow with details, 2011

[Quadrillion Btu]

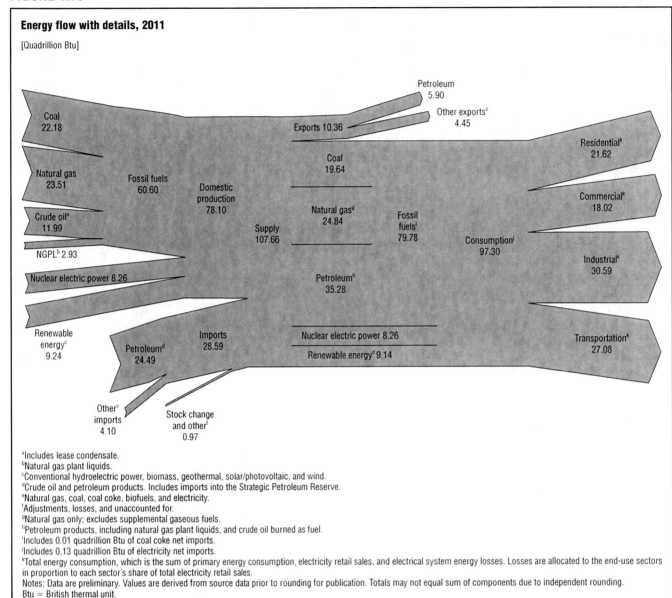

[a]Includes lease condensate.
[b]Natural gas plant liquids.
[c]Conventional hydroelectric power, biomass, geothermal, solar/photovoltaic, and wind.
[d]Crude oil and petroleum products. Includes imports into the Strategic Petroleum Reserve.
[e]Natural gas, coal, coal coke, biofuels, and electricity.
[f]Adjustments, losses, and unaccounted for.
[g]Natural gas only; excludes supplemental gaseous fuels.
[h]Petroleum products, including natural gas plant liquids, and crude oil burned as fuel.
[i]Includes 0.01 quadrillion Btu of coal coke net imports.
[j]Includes 0.13 quadrillion Btu of electricity net imports.
[k]Total energy consumption, which is the sum of primary energy consumption, electricity retail sales, and electrical system energy losses. Losses are allocated to the end-use sectors in proportion to each sector's share of total electricity retail sales.
Notes: Data are preliminary. Values are derived from source data prior to rounding for publication. Totals may not equal sum of components due to independent rounding.
Btu = British thermal unit.

SOURCE: "Figure 1.0. Energy Flow, 2011 (Quadrillion Btu)," in *Annual Energy Review 2011*, U.S. Energy Information Administration, September 27, 2012, http://www.eia.gov/totalenergy/data/annual/pdf/aer.pdf (accessed October 3, 2012)

EIA, consumer prices by energy type in 2010 were as follows:

- Retail electricity—$28.92 per million Btu
- Petroleum—$20.32 per million Btu
- Natural gas—$7.41 per million Btu
- Biomass—$3.45 per million Btu
- Coal—$2.42 per million Btu
- Nuclear fuel—$0.62 per million Btu

Overall, consumers spent $1.2 trillion on energy in 2010, which is broken down as follows:

- Petroleum—$709.5 billion
- Retail electricity—$365.9 billion
- Natural gas—$159.8 billion
- Coal—$50.4 billion
- Biomass—$6.4 billion
- Nuclear fuel—$5.2 billion

External Costs

External costs are costs other than market costs. In economics, externalities are benefits (positive consequences) and harms (negative consequences) that are not included in marketplace prices. There are positive externalities that are associated with the energy industry. For example, the research and development of energy technologies has produced scientific knowledge that has benefitted other industries, and hence the overall economy.

FIGURE 1.17

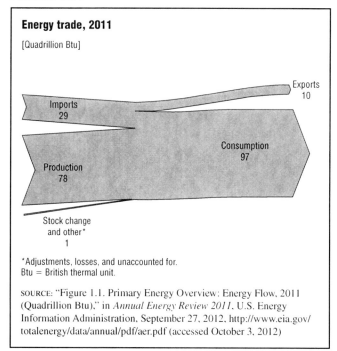

Energy trade, 2011

[Quadrillion Btu]

Imports
29

Exports
10

Production
78

Consumption
97

Stock change
and other*
1

*Adjustments, losses, and unaccounted for.
Btu = British thermal unit.

SOURCE: "Figure 1.1. Primary Energy Overview: Energy Flow, 2011
(Quadrillion Btu)," in *Annual Energy Review 2011*, U.S. Energy
Information Administration, September 27, 2012, http://www.eia.gov/
totalenergy/data/annual/pdf/aer.pdf (accessed October 3, 2012)

However, energy externalities are most often discussed in terms of negative consequences, meaning the external costs that are associated with energy. Even though these costs are believed to be substantial, they are difficult to quantify.

The most obvious external cost that is associated with energy is environmental degradation. Extracting, processing, and burning fossil fuels has environmental consequences, chiefly emissions of contaminants that degrade the air, water, and land and harm ecosystems and ultimately human health. Likewise, the production of nuclear energy and energy from renewable sources all have negative environmental impacts to varying degrees.

Of course, the energy industry (like all industries) is bound by government regulations to limit its emissions and other environmental impacts through control measures. These measures are inherently imperfect, and the imposition of stricter controls is fraught with political controversy in the United States. Energy—especially domestically produced low-cost energy—is considered to be a vital commodity for the nation's economic well-being and national security. As a result, the U.S. government supports the energy industry via many measures, which will be explained in detail later in this chapter.

Critics complain that the external environmental costs of fossil fuel use have for decades been borne by society at large, rather than by the producers and consumers of the fuels. The criticism grew especially loud during the first decade of the 21st century as energy prices soared and oil and natural gas companies made record profits. At the same time, there was growing concern about global warm-ing and resulting climate change. Scientists believe that large-scale burning of fossil fuels for more than a century has pumped enormous amounts of carbon into the atmosphere. The resulting atmospheric changes have been gradually warming the planet and changing historical climate patterns. These effects are expected to continue well into the future. Even though there are other contributing factors to global warming, including deforestation, the combustion of coal, natural gas, and oil and its derivatives (such as gasoline) is believed to be the main culprit. Thus, the negative impacts of global warming, including melting ice caps, coastal flooding, and disruptive climate changes, are considered to be external costs of fossil fuel usage.

Some people believe the federal government should force external environmental costs into the market costs for fossil fuels, such as by taxing carbon emissions from fossil fuel combustion. Advocates of this approach argue that it would show consumers the "true" costs of fossil fuels. Of course, prices would be much higher for coal, natural gas, oil, gasoline, and electricity, which is largely fossil-fuel based in the United States. Price shocks do lead to less usage of expensive commodities, and consumers tend to switch to alternatives, in this case, energy that is produced from sources other than fossil fuels, such as renewable sources like wind and solar power. A carbon tax or similar provision would certainly push the nation as a whole away from fossil fuels and toward renewables, which is a highly desirable outcome in some opinions. However, severe price shocks, especially for a widely used and vital commodity such as energy, would stress the nation's economy, perhaps on a large and calamitous scale. As a result, as of December 2012 the U.S. government had not implemented taxes that were specifically designed to capture the external environmental costs of fossil fuel use.

GOVERNMENT INTERVENTION

Government intervention has long been a factor in energy supply and demand in the United States. Over the decades presidents have set energy policies and proposed federal budgets that reflected their energy priorities. Congress has passed laws and spending bills that sometimes supported presidential priorities and sometimes reflected alternative priorities. Politics has always been a major consideration in the nation's energy decisions. At the heart of the debate is how large a role the government should play in a private market. In general, it is agreed that energy is so precious a commodity that the government should play some role, chiefly by promoting domestic production and conservation, which are both elements of the overall goal of energy self-sufficiency. This puts the government in the odd position of both encouraging domestic suppliers to bring more energy to market and at the same time encouraging buyers to use less energy. Also, because most U.S. energy is fossil-fuel based, government support of domestic fossil fuel production engenders additional environmental problems, including global warming. These seemingly contradictory outcomes are just some of the difficulties that are involved in trying to manipulate the complicated energy market.

The government uses a variety of measures to influence energy production, consumption, and costs. The main tools are tax provisions, subsidies (e.g., payments, grants, loans, and loan guarantees), resource leases, liability limits, and regulations and mandates.

Tax Provisions

The government collects taxes as a matter of course on economic activity. Energy products, like other commodities in commerce, are subject to these taxes. In addition, the government uses taxes in some areas of the energy industry to offset some of the external costs that are associated with energy development and consumption.

Taxes are sometimes imposed specifically to discourage certain activities, whereas tax breaks are granted to encourage other activities. For example, in *Fuel Economy Guide: Model Year 2012* (November 7, 2012, http://www.fueleconomy.gov/feg/pdfs/guides/FEG2012.pdf), the DOE and the U.S. Environmental Protection Agency describe two tax provisions that were active in 2012. A "gas guzzler" tax was imposed on auto manufacturers that sold cars with "exceptionally low" fuel economy. By contrast, a federal income tax credit was offered to consumers who bought certain electric or partially electric vehicles (i.e., cars using little to no gasoline). Both tax provisions directly support national goals for energy self-sufficiency.

TAX BREAKS. The United States has a long history of using tax breaks to encourage domestic energy development. Molly F. Sherlock of the Congressional Research Service notes in *Energy Tax Policy: Historical Perspectives on and Current Status of Energy Tax Expenditures* (May 2, 2011, http://www.ieeeusa.org/policy/eyeonwashington/2011/documents/energytax.pdf) that the first federal energy tax breaks were implemented in 1916. Sherlock indicates that from 1916 through 1970 tax policy "focused almost exclusively" on increasing domestic reserves and production of oil and natural gas. During the 1970s, as noted earlier, the United States suffered an energy crisis due to high reliance on foreign oil. At the same time, awareness was arising about the environmental consequences of fossil fuels. Since the 1970s, and particularly since the 1990s, the government's tax provisions have reflected growing support for energy conservation and the production and use of nonfossil fuels (i.e., so-called alternative or renewable energy sources such as solar and wind energy).

It should be noted that government intervention does not always achieve the desired results. For example, tax provisions that were favorable to renewable energy sources had little effect during the late 1990s because oil prices were historically low. Consumers responded to the low oil prices by buying large sport-utility vehicles and other vehicles that had relatively poor fuel economy. Sherlock notes that "there was little economic incentive to invest in energy efficiency, renewable energy, or renewable fuels."

Energy economics changed over the following decade as energy prices began to rise. Congress responded with the Energy Policy Act of 2005. According to Sherlock, the law "substantially" increased the number and scope of energy tax breaks, particularly for electricity infrastructure, domestic fossil fuel security, energy efficiency, alternative motor fuels and fuel economy, and energy research. As noted earlier, the Great Recession lasted from late 2007 to mid-2009. In 2008 Congress passed the Emergency Economic Stabilization Act, which extended many previous energy tax breaks, mostly to the renewable energy industries. It also raised taxes on the oil and natural gas industries to offset the lost revenues to the government from the tax breaks that were extended to the renewable energy industries.

Sherlock notes that during the first administration (2009–2013) of President Barack Obama (1961–), several bills were passed to expand and extend energy tax breaks that were devoted to renewable energy sources, energy efficiency, and alternative fuel vehicles. Obama repeatedly asked Congress to end the tax breaks that were given to the oil and natural gas industries, but as of December 2012, those tax breaks were still in effect.

Subsidies

Subsidies are direct financial incentives, such as grants, loans, or loan guarantees, that promote particular activities (e.g., energy research or conservation). The U.S. government has long used subsidies to encourage production in certain industries, particularly the agricultural industry. Energy subsidies are generally awarded by the DOE and originate from funds that are specifically devoted to that purpose in the federal budget.

Tax Breaks and Subsidies in 2011

The tax breaks and subsidies that benefit the energy industry amount to billions of dollars each year. The tax breaks represent lost revenues to the government, whereas the subsidies are expenses funded by U.S. taxpayers through the federal budget.

The Congressional Budget Office notes in *Federal Financial Support for the Development and Production of Fuels and Energy Technologies* (March 2012, http://www.cbo.gov/sites/default/files/cbofiles/attachments/03-06-

FuelsandEnergy_Brief.pdf) that in 2011 energy tax preferences and financial incentives totaled $24 billion. Tax preferences (including grants in lieu of tax credits) accounted for the vast majority of the total, at $20.5 billion. Subsidies provided by the DOE that supported energy technologies accounted for $3.5 billion of the total.

The tax preferences and grants in lieu of tax credits supported the following general energy areas:

- Renewable energy—$12.9 billion

- Fossil fuels—$2.5 billion

- Energy efficiency—$1.5 billion

- Nuclear energy—$0.9 billion

- Other—$2.7 billion

The Congressional Budget Office notes that most of the renewable energy and energy efficiency provisions expired at the end of 2011, meaning that they were only temporary measures. The provisions supporting fossil fuels and nuclear energy did not have expiration dates. This difference is emblematic of a long political struggle over what types of energy sources the government should and should not promote.

Targeted Taxes

Government entities at the federal, state, and local level often use taxes that target particular energy sectors, such as coal mining. One purpose of these targeted taxes is to offset some of the external costs that are associated with energy production and consumption, such as health, social, and environmental consequences. The collected monies may be put into trust funds that are administered by government agencies. The trust funds provide monies to cover external costs that are ongoing or may occur in the future. One example is the Oil Spill Liability Trust Fund, which was created by Congress in 1986. Its purpose is to cover certain damages that are associated with oil spills.

Government entities also commonly impose excise taxes (i.e., taxes that target specific goods) on gasoline. These taxes are often used for building and maintaining transportation infrastructure, such as roads and bridges. Many states levy severance taxes on companies that extract subsurface resources, such as coal, within their boundaries.

Whatever their purposes, taxes on energy products elevate the prices that consumers ultimately pay for those products.

Resource Leases

Another way in which the government influences energy markets is through resource leases. U.S. law provides that underground resources, such as minerals or natural gas, belong to the property owner. Thus, individuals and companies that own land can sell their underground resources to others. The federal and state governments own huge swaths of land across the United States, both onshore and offshore (i.e., off the coastlines beneath ocean waters). Government entities can sell the resources beneath the land within their jurisdictions so long as those sales do not violate existing law. As will be explained in later chapters, some government lands have been deemed off limits for resource exploration and extraction.

Resource sales by the government are carried out through legal contracts called leases. Leases cover particular tracts of land and are typically sold through auctions to the highest bidder. In addition, the leases require developers to pay the government a specific percentage of the value of any resources that are extracted. The government's control of resource leases on public lands is fraught with controversy. Policymakers, the energy industry, environmentalists, and other parties can have sharp disagreements about where the leases should be offered and how much they should cost.

Liability Limits

U.S. law includes provisions that limit the financial liability of certain energy sectors in the event of disasters within the industry. Energy accidents, such as oil spills and radiation leaks, can have enormous costs. The liability limits were set to grant the industries a level of financial protection in their exploration and production endeavors. Most of the limits were originally set decades ago. For example, the Atomic Energy Act of 1954 established liability limits on accidents that were associated with the U.S. nuclear power industry. Lawmakers feared that without the limits, the fledgling industry would not grow because companies would not be able to afford the premiums for insurance policies protecting them against damages. In addition, the limits have served to keep energy prices lower because they represent a cost savings to energy producers. However, in the 21st century the limits have become quite controversial because they are seen as relatively low in comparison to the massive costs that can arise from an energy-related disaster. Liability limits specific to various energy sectors are described in subsequent chapters.

Regulations and Mandates

The government uses regulations and mandates to exert control over energy production and consumption factors. For example, the energy industry is subject to extensive environmental regulations that limit the types and amounts of emissions that energy producers can release to the air, land, and water. These regulations are designed to minimize the external costs that are associated with energy production, but their imposition does increase the market costs of energy sources.

Mandates are usually directed at supporting a particular type of energy source. Chapter 6 describes federal and state government mandates for use of specified levels of biomass and other renewable energy sources. For example, California requires that renewable energy sources account for 33% of the electricity sales in the state by 2020.

THE DOMESTIC OUTLOOK

In *Annual Energy Outlook 2012* (June 2012, http://www.eia.gov/forecasts/aeo/pdf/0383(2012).pdf), the EIA forecasts energy supply, demand, and prices through 2035 for various scenarios based on expected supply and demand factors. Table 1.4 summarizes key information known for 2010 and projected for 2035 for a reference case that assumes future energy markets and policies will be similar to those currently in place.

As shown in Table 1.4, total domestic energy production is projected to increase from 75.5 quadrillion Btu in 2010 to 94.7 quadrillion Btu in 2035, or 0.9% annually. Total energy imports are forecast to decrease by 0.7% annually, whereas energy exports are forecast to increase by 1.7% annually. Overall, domestic consumption is expected to increase from 98.2 quadrillion Btu in 2010 to 106.9 quadrillion Btu in 2035, or 0.3% annually. Prices are shown in Table 1.4 for two price types: nominal and real. Nominal prices are not adjusted for inflation; they represent the actual dollar amounts that consumers pay at specific times. Real prices are adjusted for inflation—that is, they assume the dollar has the same value over time, in this case the value it had in 2010. Comparison of real prices between 2010 and 2035 shows the effects on energy prices solely by market factors. Real prices for all energy commodities are expected to rise by 2035. Petroleum shows the largest annual price increase, whereas electricity shows the lowest increase. Table 1.5 shows more detailed information for real price changes that are forecast for the energy-using sectors by energy source. Overall, petroleum products, such as fuel oils and jet fuel, are projected to have the largest annual price increases, particularly for the electric power, transportation, and industrial sectors.

WORLD ENERGY PRODUCTION AND CONSUMPTION

Table 1.6 shows world energy production for 1980, 1990, 2000, and 2009 for the world, by region, and for the top-25 producers. The total world production increased by 68% between 1980 and 2009, from 287.3 quadrillion Btu to 484.2 quadrillion Btu. Twenty-three percent of that growth occurred between 2000 and 2009. The Asia and Oceania region experienced the largest growth, by far, between 2000 and 2009, boosting its energy production by 83%. The United States was

TABLE 1.4

Total energy supply, disposition, and price summary, 2010 and 2035

[Quadrillion Btu per year, unless otherwise noted]

Supply, disposition, and prices	Reference case 2010	Reference case 2035	Annual growth 2010–2035 (percent)
Production			
Crude oil and lease condensate	11.59	12.89	0.4%
Natural gas plant liquids	2.78	3.94	1.4%
Dry natural gas	22.10	28.60	1.0%
Coal[a]	22.06	24.14	0.4%
Nuclear/uranium[b]	8.44	9.28	0.4%
Hydropower	2.51	3.04	0.8%
Biomass	4.05	9.07	3.3%
Other renewable energy	1.34	2.81	3.0%
Other[5]	0.64	0.91	1.4%
Total	**75.50**	**94.67**	**0.9%**
Imports			
Crude oil	20.14	16.90	−0.7%
Liquid fuels and other petroleum[d]	5.02	4.14	−0.8%
Natural gas[e]	3.81	2.84	−1.2%
Other imports[f]	0.52	0.81	1.8%
Total	**29.49**	**24.69**	**−0.7%**
Exports			
Liquid fuels and other petroleum[g]	4.81	4.95	0.1%
Natural gas[h]	1.15	4.17	5.3%
Coal	2.10	3.13	1.6%
Total	**8.06**	**12.25**	**1.7%**
Discrepancy[i]	**−1.23**	**0.18**	**—**
Consumption			
Liquid fuels and other petroleum	37.25	37.70	0.0%
Natural gas	24.71	27.26	0.4%
Coal[j]	20.76	21.15	0.1%
Nuclear/uranium[f]	8.44	9.28	0.4%
Hydropower	2.51	3.04	0.8%
Biomass[k]	2.88	5.44	2.6%
Other renewable energy	1.34	2.81	3.0%
Other[l]	0.29	0.24	−0.6%
Total	**98.16**	**106.93**	**0.3%**
Prices (2010 dollars per unit)			
Petroleum (dollars per barrel)			
Low sulfur light crude oil	79.39	144.98	2.4%
Imported crude oil[m]	75.87	132.95	2.3%
Natural gas (dollars per million Btu)			
at Henry hub	4.39	7.37	2.1%
at the wellhead[n]	4.06	6.48	1.9%
Natural gas (dollars per thousand cubic feet) at the wellhead[n]	4.16	6.64	1.9%
Coal (dollars per ton) at the minemouth[o]	35.61	50.52	1.4%
Coal (dollars per million Btu)			
at the minemouth	1.76	2.56	1.5%
Average end-use[p]	2.38	2.94	0.9%
Average electricity (cents per kilowatthour)	9.80	10.10	0.1%
Prices (nominal dollars per unit)			
Petroleum (dollars per barrel)			
Low sulfur light crude oil	79.39	229.55	4.3%
Imported crude oil[m]	75.87	210.51	4.2%
Natural gas (dollars per million Btu)			
at Henry hub	4.39	11.67	4.0%
at the wellhead[n]	4.06	10.26	3.8%
Natural gas (dollars per thousand cubic feet) at the wellhead[n]	4.16	10.51	3.8%
Coal (dollars per ton) at the minemouth[o]	35.61	80.00	3.3%

TABLE 1.4

Total energy supply, disposition, and price summary, 2010 and 2035 [CONTINUED]

[Quadrillion Btu per year, unless otherwise noted]

Supply, disposition, and prices	Reference case 2010	Reference case 2035	Annual growth 2010–2035 (percent)
Coal (dollars per million Btu)			
at the minemouth°	1.76	4.05	3.4%
Average end-use°	2.38	4.66	2.7%
Average electricity (cents per kilowatthour)	9.80	16.00	2.0%

°Includes waste coal.
°These values represent the energy obtained from uranium when it is used in light water reactors. The total energy content of uranium is much larger, but alternative processes are required to take advantage of it.
°Includes non-biogenic municipal waste, liquid hydrogen, methanol, and some domestic inputs to refineries.
°Includes imports of finished petroleum products, unfinished oils, alcohols, ethers, blending components, and renewable fuels such as ethanol.
°Includes imports of liquefied natural gas that is later re-exported.
°Includes coal, coal coke (net), and electricity (net). Excludes imports of fuel used in nuclear power plants.
°Includes crude oil, petroleum products, ethanol, and biodiesel.
°Includes re-exported liquefied natural gas.
°Balancing item. Includes unaccounted for supply, losses, gains, and net storage withdrawals.
°Excludes coal converted to coal-based synthetic liquids and natural gas.
°Includes grid-connected electricity from wood and wood waste, non-electric energy from wood, and biofuels heat and coproducts used in the production of liquid fuels, but excludes the energy content of the liquid fuels.
°Includes non-biogenic municipal waste, liquid hydrogen, and net electricity imports.
°Weighted average price delivered to U.S. refiners.
°Represents lower 48 onshore and offshore supplies.
°Includes reported prices for both open market and captive mines.
°Prices weighted by consumption; weighted average excludes residential and commercial prices, and export free-alongside-ship (f.a.s.) prices.
Btu = British thermal unit.
— = Not applicable.
Note: Totals may not equal sum of components due to independent rounding. Data for 2009 and 2010 are model results and may differ slightly from official EIA data reports.

SOURCE: Adapted from "Table A1. Total Energy Supply, Disposition, and Price Summary," in *Annual Energy Outlook 2012*, U.S. Energy Information Administration, June 25, 2012, http://www.eia.gov/forecasts/aeo/pdf/0383(2012).pdf (accessed October 3, 2012

TABLE 1.5

Energy prices by sector and source, 2010 and 2035

[2010 dollars per million Btu, unless otherwise noted]

Sector and source	Reference case 2010	Reference case 2035	Annual growth 2010–2035 (percent)
Residential			
Liquefied petroleum gases	27.02	34.64	1.0%
Distillate fuel oil	21.21	32.73	1.8%
Natural gas	11.08	13.98	0.9%
Electricity	33.69	34.58	0.1%
Commercial			
Liquefied petroleum gases	23.52	31.30	1.1%
Distillate fuel oil	20.77	29.18	1.4%
Residual fuel oil	11.07	18.90	2.2%
Natural gas	9.10	11.64	1.0%
Electricity	29.73	29.48	−0.0%
Industrial[a]			
Liquefied petroleum gases	21.80	32.18	1.6%
Distillate fuel oil	21.32	29.53	1.3%
Residual fuel oil	10.92	21.65	2.8%
Natural gas[b]	5.51	7.54	1.3%
Metallurgical coal	5.84	9.11	1.8%
Other industrial coal	2.71	3.64	1.2%
Coal to liquids	—	2.38	—
Electricity	19.63	20.78	0.2%
Transportation			
Liquefied petroleum gases[c]	26.88	35.74	1.1%
E85[d]	25.21	31.96	1.0%
Motor gasoline[e]	22.70	33.61	1.6%
Jet fuel[f]	16.22	29.13	2.4%
Diesel fuel (distillate fuel oil)[g]	21.87	32.40	1.6%
Residual fuel oil	10.42	20.95	2.8%
Natural gas[h]	13.20	14.51	0.4%
Electricity	32.99	33.82	0.1%
Electric power[i]			
Distillate fuel oil	18.73	27.80	1.6%
Residual fuel oil	11.89	25.72	3.1%
Natural gas	5.14	7.21	1.4%
Steam coal	2.26	2.80	0.9%
Average price to all users[i]			
Liquefied petroleum gases	17.28	26.63	1.7%
E85[d]	25.21	31.96	1.0%
Motor gasoline[e]	22.59	33.61	1.6%
Jet fuel	16.22	29.13	2.4%
Distillate fuel oil	21.65	31.91	1.6%
Residual fuel oil	10.82	21.68	2.8%
Natural gas	7.16	9.30	1.1%
Metallurgical coal	5.84	9.11	1.8%
Other coal	2.29	2.85	0.9%
Coal to liquids	—	2.38	—
Electricity	28.68	29.56	0.1%

[a]Includes energy for combined heat and power plants, except those whose primary business is to sell electricity, or electricity and heat, to the public.
[b]Excludes use for lease and plant fuel.
[c]Includes federal and state taxes while excluding county and local taxes.
[d]E85 refers to a blend of 85 percent ethanol (renewable) and 15 percent motor gasoline (nonrenewable). To address cold starting issues, the percentage of ethanol varies seasonally. The annual average ethanol content of 74 percent is used for this forecast.
[e]Sales weighted-average price for all grades. Includes federal, state and local taxes.
[f]Kerosene-type jet fuel. Includes federal and state taxes while excluding county and local taxes.
[g]Diesel fuel for on-road use. Includes federal and state taxes while excluding county and local taxes.
[h]Natural gas used as a vehicle fuel. Includes estimated motor vehicle fuel taxes and estimated dispensing costs or charges.
[i]Includes electricity-only and combined heat and power plants whose primary business is to sell electricity, or electricity and heat, to the public.
[j]Weighted averages of end-use fuel prices are derived from the prices shown in each sector and the corresponding sectoral consumption.

historically the highest producer, but it was surpassed in 2009 by China, which saw a stunning 139% rise in its energy production between 2000 and 2009. Other top producers in 2009 were Russia, Saudi Arabia, and Canada. U.S. production of 72.7 quadrillion Btu in 2009 accounted for 15% of the total world production.

The total world energy consumption is shown in Table 1.7. It grew from 283.1 quadrillion Btu in 1980 to 483 quadrillion Btu in 2009, a 71% increase. Between 2000 and 2009 consumption grew by 22%. The Asia and Oceania region and Middle East region experienced the highest growth surges of 68% and 60%, respectively, between 2000 and 2009. However, the United States has historically been the leading energy consumer among countries. U.S. consumption of 94.6 quadrillion Btu in 2009 accounted for 20% of the total world consumption. China was a close second; its consumption of 90.3 quadrillion Btu was 19% of the total. Other top consumers in 2009 included Russia, India, and Japan.

TABLE 1.5

Energy prices by sector and source, 2010 and 2035 [CONTINUED]

[2010 dollars per million Btu, unless otherwise noted]

Btu = British thermal unit.
— = Not applicable.
Note: Data for 2009 and 2010 are model results and may differ slightly from offcial EIA data reports.

SOURCE: Adapted from "Table A3. Energy Prices by Sector and Source," in *Annual Energy Outlook 2012*, U.S. Energy Information Administration, June 25, 2012, http://www.eia.gov/forecasts/aeo/pdf/0383(2012).pdf (accessed October 3, 2012)

TABLE 1.6

World primary energy production, by region and selected country, selected years, 1980–2009

[Quadrillion Btus]

	1980	1990	2000	2009	% change between 2000 and 2009
World	287.34	349.51	392.39	484.15	23%
Asia & Oceania	35.80	59.49	77.80	142.10	83%
North America	83.18	91.77	98.83	99.50	1%
Middle East	42.26	40.99	57.48	66.23	15%
Eurasia	56.46	72.11	54.13	66.07	22%
Europe	40.17	46.85	50.51	44.65	−12%
Africa	17.40	21.63	27.69	36.13	30%
Central & South America	12.06	16.68	25.97	29.47	13%
China	18.12	29.39	34.20	81.89	139%
United States	67.18	70.70	71.33	72.65	2%
Russia	—	—	41.69	49.51	19%
Saudi Arabia	22.43	15.92	21.59	22.91	6%
Canada	10.28	13.40	18.12	18.25	1%
India	3.10	6.82	9.83	14.58	48%
Iran	3.94	7.67	10.40	14.26	37%
Indonesia	4.23	5.30	7.68	12.71	65%
Australia	3.25	6.16	9.66	12.26	27%
Norway	2.93	5.80	10.27	9.89	−4%
Brazil	1.90	3.76	6.35	8.91	40%
Mexico	5.72	7.66	9.37	8.61	−8%
United Arab Emirates	3.89	5.51	6.77	7.52	11%
Venezuela	5.77	6.31	9.37	7.22	−23%
Algeria	2.80	4.75	6.19	7.06	14%
United Kingdom	8.66	9.01	11.08	6.66	−40%
South Africa	2.72	4.05	5.58	6.09	9%
Nigeria	4.50	4.07	5.18	5.74	11%
Kuwait	3.99	2.83	5.04	5.73	14%
Qatar	1.20	1.20	2.82	5.62	99%
Kazakhstan	—	—	3.38	5.56	64%
Iraq	5.45	4.54	5.62	5.15	−8%
France	2.33	4.25	5.02	4.78	−5%
Germany	—	—	5.30	4.71	−11%
Libya	4.03	3.18	3.30	4.29	30%

— = Not applicable
Btu = British thermal unit.

SOURCE: Adapted from "Table. Total Primary Energy Production (Quadrillion Btu)," in *International Energy Statistics*, U.S. Energy Information Administration, 2012, http://www.eia.gov/cfapps/ipdbproject/IEDIndex3.cfm?tid=44&pid=44&aid=1 (accessed October 5, 2012)

TABLE 1.7

World primary energy consumption, by region and selected country, selected years, 1980–2009

[Quadrillion Btus]

	1980	1990	2000	2009	% change between 2000 and 2009
World	283.05	346.93	394.99	482.97	22%
Asia & Oceania	48.90	74.39	106.24	178.03	68%
North America	91.50	100.14	118.22	114.60	−3%
Europe	71.75	76.26	81.17	81.05	−0.1%
Eurasia	46.74	60.98	39.19	40.20	3%
Middle East	5.84	11.22	17.35	27.79	60%
Central & South America	11.54	14.48	20.79	25.25	21%
Africa	6.80	9.46	12.03	16.05	33%
United States	78.07	84.49	98.81	94.55	−4%
China	17.29	27.00	36.35	90.26	148%
Russia	—	—	26.14	26.82	3%
India	4.04	7.88	13.46	21.69	61%
Japan	15.20	18.76	22.40	20.60	−8%
Germany	—	—	14.24	13.46	−6%
Canada	9.69	10.98	13.06	13.04	−0.2%
France	8.39	9.13	10.85	10.66	−2%
Brazil	4.02	5.75	8.53	10.29	21%
Korea, South	1.76	3.84	7.84	9.96	27%
Iran	1.58	3.10	5.02	9.02	79%
United Kingdom	8.84	9.28	9.73	8.91	−8%
Saudi Arabia	1.66	3.35	4.85	7.83	61%
Italy	6.12	6.72	7.59	7.32	−4%
Mexico	3.72	4.65	6.33	7.00	10%
Spain	3.04	3.92	5.53	6.08	10%
Indonesia	1.17	2.31	3.87	6.05	56%
Australia	2.74	3.71	4.83	5.59	16%
South Africa	2.73	3.52	4.58	5.47	19%
Ukraine	—	—	5.75	4.70	−18%
Taiwan	1.12	2.03	3.81	4.49	18%
Netherlands	3.21	3.31	3.79	4.05	7%
Turkey	1.04	1.97	3.16	4.04	28%
Thailand	0.51	1.25	2.58	4.00	55%
Poland	5.07	3.95	3.62	3.98	10%

— = Not applicable
Btu = British thermal unit.

SOURCE: Adapted from "Table. Total Primary Energy Consumption (Quadrillion Btu)," in *International Energy Statistics*, U.S. Energy Information Administration, 2012, http://www.eia.gov/cfapps/ipdbproject/IEDIndex3.cfm?tid=44&pid=44&aid=2 (accessed October 5, 2012)

CHAPTER 2
OIL

On August 27, 1859, Edwin Drake (1819–1880) struck oil 69 feet (21 m) below the surface of the earth near Titusville, Pennsylvania. This was the first successful modern oil well and ushered in a new wave of modernization. Not only did oil help meet the growing demand for new and better fuels for heating and lighting but also it proved to be an excellent source of gasoline for the internal combustion engine, which was developed during the late 1800s. Oil has become one of the most valuable commodities on the earth. Its supply is finite, and the United States consumes more oil every year than any other country, primarily because of Americans' love of the automobile. This addiction has its price; the United States has been forced for decades to import oil from countries that are sometimes less than friendly, sometimes downright hostile. Thus, oil self-sufficiency is a major goal for the United States.

UNDERSTANDING OIL

Oil is a generic term for liquid fossil fuels. As explained in Chapter 1, fossil fuels are the below-ground remains of prehistoric organisms that have become energy enriched after millions of years of being exposed to high temperatures and pressures. Oil and natural gas are often found together because they both originate from microscopic plants and animals that died in ancient water bodies (mostly swamps) and were gradually buried under deeper and deeper layers of sediment. Over several millennia the layers were compressed and baked into rock beds, and the original microorganisms were chemically transformed into hydrocarbon chemicals. Crude oil forms at slightly shallower depths and lower temperatures than does natural gas. (See Figure 2.1.)

Both substances are less dense than their source rock and once formed, they begin moving through the pores and fractures of surrounding rock formations. They migrate upward toward areas of lower pressure until a nonporous nonfractured rock layer stops them or they seep from the ground into the open air. Underground oil and natural gas accumulations are called reservoirs, fields, or pools. They are typically found in sandstone or limestone formations that are overlaid with a shield layer of impermeable rock or shale. The oil and natural gas are trapped beneath the shield within the numerous pores and fractures of the reservoir rock. Because gas is lighter than liquid, the natural gas accumulates at the top of the traps above the oil pools. Anticlines (archlike folds in a bed of rock), faults, and salt domes are common trapping formations. (See Figure 2.2.) Oil fields can be found at varying depths below the ground surface. Wells are drilled to reach the reservoirs and extract the oil.

It should be noted that the previous description refers to crude oil. In *Annual Energy Review 2011* (September 2012, http://www.eia.gov/totalenergy/data/annual/pdf/aer.pdf), the Energy Information Administration (EIA) within the U.S. Department of Energy defines crude oil as "a mixture of hydrocarbons that exists in liquid phase in natural underground reservoirs and remains liquid at atmospheric pressure after passing through surface separating facilities." Crude oil is known as conventional oil because for more than a century it has been the preferred and most commercialized type of oil. However, there are other types of naturally occurring hydrocarbon oils. These so-called unconventional oils come from two different underground formations: oil shale and tar sands (or oil sands).

The U.S. Department of the Interior's Bureau of Land Management (BLM) notes in "Frequently Asked Questions" (2012, http://ostseis.anl.gov/faq/index.cfm) that oil shale "generally refers to any sedimentary rock that contains solid bituminous materials ... that are released as petroleum-like liquids when the rock is heated in the chemical process of pyrolysis." These rocks contain kerogen, a waxy material that is a precursor to crude oil.

FIGURE 2.1

Burial temperature effects on hydrocarbon transformation

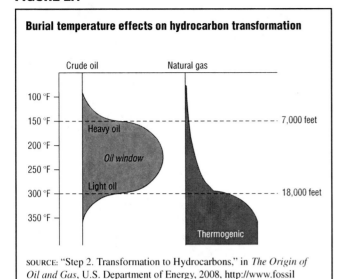

SOURCE: "Step 2. Transformation to Hydrocarbons," in *The Origin of Oil and Gas*, U.S. Department of Energy, 2008, http://www.fossil.energy.gov/programs/reserves/npr/publications/The_Origin_of_Oil_and_Gas1.pdf (accessed October 5, 2012)

FIGURE 2.2

Petroleum traps

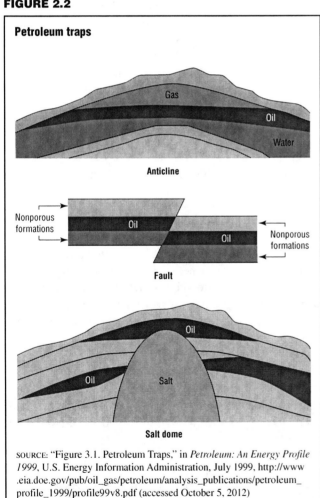

SOURCE: "Figure 3.1. Petroleum Traps," in *Petroleum: An Energy Profile 1999*, U.S. Energy Information Administration, July 1999, http://www.eia.doe.gov/pub/oil_gas/petroleum/analysis_publications/petroleum_profile_1999/profile99v8.pdf (accessed October 5, 2012)

Tar sands are "a combination of clay, sand, water, and bitumen." Bitumen is a black tarry substance that is also a precursor to crude oil. Thus, oil shale and tar sands contain oily deposits that have not yet been completely

transformed to crude oil. The deposits are said to be "geo-thermally immature." Technological advances over the past half-century have allowed engineers to liberate and liquefy these unconventional oils. They can then be processed and refined along with crude oil. As will be explained later, the extraction of unconventional oils is far more difficult than that of crude oil.

The generic term *petroleum* is often used as a synonym for oil. The word *petroleum* is derived from the Latin words "petra" (meaning rock) and "oleum" (meaning oil). Thus, petroleum originally meant "rock oil," a very apt description for the fossil fuel. The EIA defines petroleum as a "class of liquid hydrocarbon mixtures," including crude oil and its derivatives, and other liquids, particularly liquids that are associated with natural gas extraction. The latter include lease condensate and natural gas plant liquids. Lease condensate is a liquid recovered from natural gas at the well (the extraction point). It consists primarily of chemical compounds called pentanes and heavier hydrocarbons and is generally blended with crude oil for refining. Natural gas plant liquids, such as butane and propane, are recovered during the refinement of natural gas in processing plants.

Oil Deposits

The locations of oil deposits around the world are driven solely by geology. Regions and countries that are known to have large oil fields include the Middle East, Russia, China, northern South America, Mexico, Canada, and the United States.

Many of the earth's oil fields are offshore (i.e., they lie beneath the oceans). The EIA explains in "Offshore Oil and Gas" (July 3, 2012, http://www.eia.gov/energyexplained/index.cfm?page=oil_offshore) that the United States lays claim to an Exclusive Economic Zone (EEZ) that extends outward 200 miles (322 km) from its coastline. This applies to the U.S. mainland, Alaska, Hawaii, and all U.S. territories. (See Figure 2.3.) Figure 2.4 shows the extent of the EEZ and a conceptual side view of a continental shelf. The latter is the "shelf" of underwater land that extends outward from the coastline of each continent. The outward extent of the North American continental shelf ranges from around 12 miles (20 km) to 250 miles (400 km). According to the EIA, the water depth above this continental shelf is typically less than 492 feet (150 m) to 656 feet (200 m) at its deepest point. The continental shelf ends with a sharp drop off to the continental slope, which underlies deep ocean. (See Figure 2.4.) The United States claims exclusive rights to the resources (such as oil) underlying the continental shelf off its coastlines.

In general, the continental shelf is relatively narrow along the Pacific coast, wide along much of the Atlantic coast and the Gulf of Alaska, and widest in the Gulf of Mexico. Most coastal states have resource rights extending

FIGURE 2.3

Map of U.S. Exclusive Economic Zone

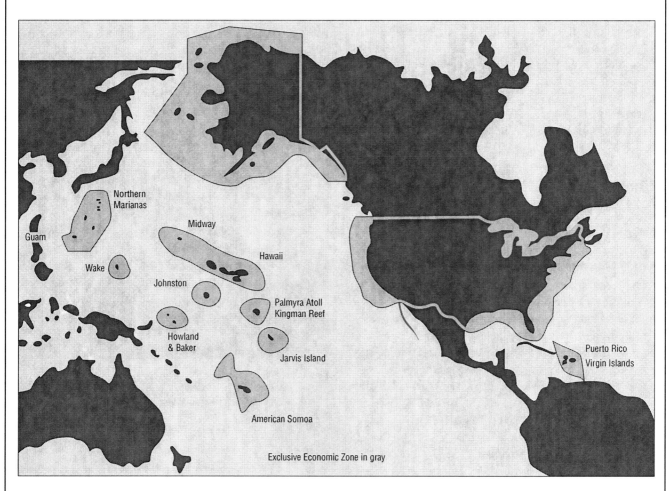

Northern Marianas

Guam

Wake

Midway

Hawaii

Johnston

Palmyra Atoll
Kingman Reef

Howland
& Baker

Jarvis Island

Puerto Rico
Virgin Islands

American Somoa

Exclusive Economic Zone in gray

SOURCE: "Map Showing Exclusive Economic Zone around the United States and Territories," in *Oil: Crude and Petroleum Products Explained: Offshore Oil and Gas*, U.S. Energy Information Administration, July 3, 2012, http://www.eia.gov/energyexplained/images/charts/EEZmap.png (accessed October 7, 2012)

FIGURE 2.4

Offshore side view

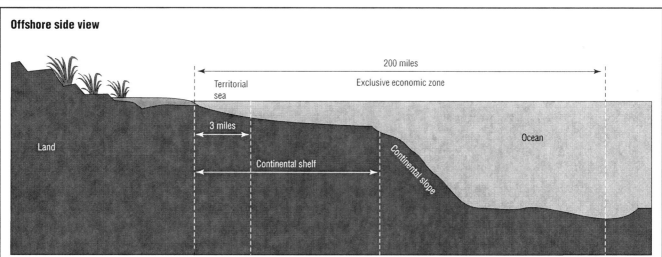

200 miles

Territorial
sea

Exclusive economic zone

3 miles

Continental slope

Ocean

Land

Continental shelf

SOURCE: "Diagram of Shore and Ocean Overlaid with Territorial Sea, Exclusive Economic Zone, the Continental Shelf, and Continental Slope," in *Oil: Crude and Petroleum Products Explained: Offshore Oil and Gas*, U.S. Energy Information Administration, July 3, 2012, http://www.eia.gov/energyexplained/images/continentalshelf.gif (accessed October 7, 2012)

outward 3 miles (5 km) from their coastlines. Resources beyond that limit fall under federal government control in an area called the outer continental shelf (OCS). Water above the OCS can be up to 600 feet (183 m) deep.

Estimates of the total amounts of oil deposits around the world and in the United States that have not yet been extracted are provided in Chapter 7, along with information about oil exploration and development activities.

Measuring Oil

In the United States raw (unprocessed) oil volumes are measured in barrels. A barrel is equivalent to 42 gallons (159 L). Because a barrel is a relatively small unit of measure, oil data are typically expressed in units of thousand barrels or million barrels. A commonly used unit for production and consumption data is million barrels per day (mbpd).

Grading Oil Quality

Oil from different fields can differ substantially in terms of physical and chemical properties. Because crude oil, shale oil, and tar sands oil are all refined before being used, their characteristics influence the technologies and costs that are associated with refining them. Some key properties include density, viscosity (thickness, or resistance to flow), heat content, and the amounts of water, minerals, combustible gases, and sulfur. Oil's viscosity can vary considerably. Oil that flows easily (like water) is said to be "light," and thick dense oil is said to be "heavy." In addition, oil is called "sweet" if it contains only a small amount of sulfur and "sour" if it contains a lot of sulfur. Refiners prefer sweet light oil because it is easy to pump and does not require extensive treatment for sulfur removal.

OIL EXTRACTION

Crude Oil

Most crude oil wells are drilled with a rotary drilling system, or rotary rig, as illustrated in Figure 2.5. The rotating bit at the end of a pipe drills a hole into the ground. Drilling mud is pushed down through the pipe and the drill bit, forcing small pieces of drilled rock to the surface, as shown by the arrows in the diagram. As the well gets deeper, more pipe is added. The oil derrick above the ground supports equipment that can lift the pipe and drill bit from the well when drill bits need to be changed or replaced.

Drilling takes place both onshore and offshore. The development of offshore oil and gas resources began with the drilling of the Summerland oil field along the coast of California in 1896, where about 400 wells were drilled. Since then, the industry has continually improved drilling technology. In the 21st century deepwater petroleum and natural gas exploration occurs from platforms and drill

FIGURE 2.5

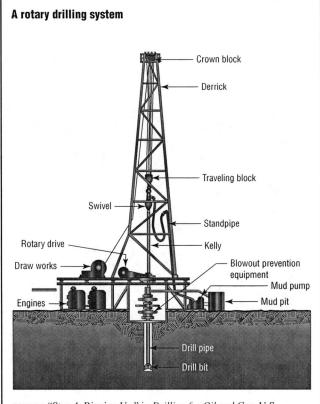

A rotary drilling system

SOURCE: "Step 4. Rigging Up," in *Drilling for Oil and Gas*, U.S. Department of Energy, 2008, http://www.fossil.energy.gov/programs/reserves/npr/publications/Drilling_for_Oil_and_Gas.pdf (accessed October 5, 2012)

ships, and shallow-water exploration occurs from gravel islands (humanmade islands of gravel and sand) and mobile units (ship-based drilling units).

After crude oil reservoirs have been tapped for several years or decades, their supply of oil becomes depleted. Several techniques can be used to recover additional oil, including the injection of water, chemicals, or steam to force more oil from the rock. These recovery techniques can be expensive and add to the cost of producing each barrel of crude oil.

Unconventional Oils

As noted earlier, oil shale and tar sands contain oil deposits called kerogen and bitumen, respectively. Because these deposits are solid or semisolid, they cannot be pumped out of the ground like crude oil. Instead, they are typically dug out using mining techniques. In "About Oil Shale" (2012, http://ostseis.anl.gov/guide/oilshale/index.cfm), the BLM notes that mined kerogen is retorted (heated to a high temperature) at the surface to liquefy the oil for separation and collection.

Tar sands that are near the surface of the earth are recovered by open-pit mining, a process in which huge shovels and trucks dig it up after removing the soil, clay, and gravel that lie on top. Tar sands that are deep within the earth are recovered by drilling to the deposit and then injecting steam to bring bitumen to the surface. Other methods are sometimes used as well, but all efforts amount to an extremely expensive and specialized process.

Because the oil extracted from oil shale and tar sands is typically blended into crude oil, many analysts refer to all "raw" oil as crude oil. This naming convention will be followed in this book when referring to oil past the extraction stage.

CRUDE OIL TRANSPORT

According to the EIA, in "Petroleum and Other Liquids" (September 27, 2012, http://www.eia.gov/dnav /pet/pet_move_imp_dc_NUS-Z00_mbbl_a.htm), nearly 3.3 billion barrels of crude oil were imported to the United States in 2011. Much of this oil traveled by sea in enormous oil tankers. In addition, there are more than a dozen oil pipelines connecting the United States and Canada. Two of the largest Canadian companies involved in piping crude oil to the United States are Enbridge Inc. (2012, http://www.enbridge.com/DeliveringEnergy/Our Pipelines/LiquidsPipelines.aspx) and TransCanada Corporation (2012, http://www.transcanada.com/100.html).

The United States features a massive, mostly interconnected network of pipelines for moving crude oil around the country. (For a map of the pipelines, see the American Petroleum Institute in "U.S. Refineries, Crude Oil and Refined Products Pipelines" [2012, http://www.api.org/oil-and-natural-gas-overview/transporting-oil-and-natural-gas/ pipeline/~/media/Files/Oil-and-Natural-Gas/pipeline/Liquid-Petroleum-Products-map.ashx].)

The Keystone XL Project

In 2010 TransCanada sought permission from the U.S. government to construct a new crude oil pipeline from Canada into the United States. The project is named Keystone XL (http://www.transcanada.com/keystone.html) and is proposed to run from Alberta, Canada, cross the U.S. border into Montana and connect in Steele City, Nebraska, with an existing TransCanada pipeline called the Keystone pipeline. The latter ends in Cushing, Oklahoma, which is a key hub for U.S. oil resources. The Keystone XL project would include an extension from Cushing to the Gulf Coast. In addition, the Keystone XL portion in Nebraska would be extended eastward into southern Illinois. The Keystone XL project has encountered stiff resistance from environmentalists and is opposed, in part, by President Barack Obama (1961–). Presidential permission is required for a pipeline to cross the U.S. border.

In "Keystone Pipeline: Separating Reality from Rhetoric" (CNNMoney.com, March 22, 2012), Chris Isidore reports that in early 2012 Obama announced that his administration would not oppose construction of the Cushing–to–Gulf Coast leg of the project. However, as Isidore points out, this section does not require presidential approval because it does not involve an international border crossing. Thus, as of December 2012, one leg of the Keystone XL project was moving forward, but the remainder had yet to gain approval.

OIL REFINING

Crude oil has little practical use in its raw state. It is a mixture of different liquid hydrocarbons that are separable by distillation. (See Figure 2.6.) The liquid components (or fractions) vaporize at different boiling points. Light fractions, such as butane, boil off at relatively low temperatures. Heavy components with high boiling points, such as residual fuel oil, must be heated to high temperatures to boil off. All the separated vapors are then cooled to condense them back to liquids. The heavier fractions are often subjected to additional refining processes called cracking and reforming to break up heavy petroleum products into lighter ones. Refining produces numerous petroleum products, all of which have different physical and chemical properties. These products include gasoline, diesel fuel, jet fuel, and lubricants for transportation; heating oil, residual oil, and kerosene for heat;

FIGURE 2.6

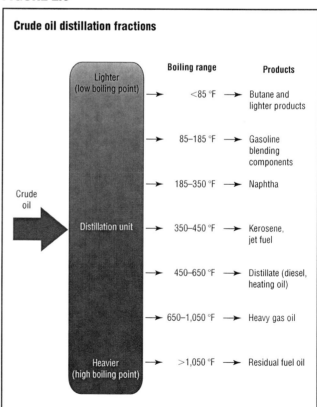

Crude oil distillation fractions

SOURCE: "Crude Oil Distillation Unit and Products," in *Today in Energy: Crude Oil Distillation and the Definition of Refinery Capacity,* U.S. Energy Information Administration, July 5, 2012, http://www.eia.gov/ todayinenergy/detail.cfm?id=6970# (accessed October 5, 2012)

butane, ethane, and propane; and heavy residuals for paving and roofing. Petroleum by-products are also vital to the chemical industry, ending up in many different foams, plastics, synthetic fabrics, paints, dyes, inks, and even pharmaceutical drugs. Because of their dependence on petroleum, many chemical plants are directly connected by pipelines to nearby refineries.

Refining is a continuous process, with raw oil entering the refinery at the same time that finished products leave by pipeline, truck, and train. At most refineries storage capacity is limited, so if there is a malfunction and products cannot be refined, the oil may be burned off (flared) rather than stored. Whereas a small flare is typical at a refinery or a chemical plant, large flares or many flares likely indicate a processing problem.

REFINERY NUMBERS AND CAPACITY

In 2011 there were 148 refineries operating in the United States, which was a drop from 336 in 1949 and 324 in 1981. (See Table 2.1.) Refinery capacity in 2011 was about 17.7 mbpd, below the 1981 peak of 18.6 mbpd. As of 2011, U.S. refineries were operating at a high capacity. Utilization rates generally increased from a low of 68.6% in 1981 (a period of low demand because

TABLE 2.1

Refinery statistics, selected years, 1949–2011

Year	Operable refineries[a] Number	Operable refineries capacity On January 1	Operable refineries capacity Annual average[b]	Gross input to distillation units	Utilization[c]
		Thousand barrels per calendar day		Thousand barrels per day	Percent
1949	336	6,231	NA	5,556	89.2
1950	320	6,223	NA	5,980	92.5
1955	296	8,386	NA	7,820	92.2
1960	309	9,843	NA	8,439	85.1
1965	293	10,420	NA	9,557	91.8
1970	276	12,021	NA	11,517	92.6
1975	279	14,961	NA	12,902	85.5
1976	276	15,237	NA	13,884	87.8
1977	282	16,398	NA	14,982	89.6
1978	296	17,048	NA	15,071	87.4
1979	308	17,441	NA	14,955	84.4
1980	319	17,988	NA	13,796	75.4
1981	324	18,621	18,603	12,752	68.6
1982	301	17,890	17,432	12,172	69.9
1983	258	16,859	16,668	11,947	71.7
1984	247	16,137	16,035	12,216	76.2
1985	223	15,659	15,671	12,165	77.6
1986	216	15,459	15,459	12,826	82.9
1987	219	15,566	15,642	13,003	83.1
1988	213	15,915	15,927	13,447	[R]84.4
1989	204	15,655	15,701	13,551	[R]86.3
1990	205	15,572	15,623	13,610	87.1
1991	202	15,676	15,707	13,508	86.0
1992	199	15,696	15,460	13,600	87.9
1993	187	15,121	15,143	13,851	91.5
1994	179	15,034	15,150	14,032	92.6
1995	175	15,434	15,346	14,119	92.0
1996	170	15,333	15,239	14,337	94.1
1997	164	15,452	15,594	14,838	95.2
1998	163	15,711	15,802	15,113	95.6
1999	159	16,261	16,282	15,080	92.6
2000	158	16,512	16,525	15,299	92.6
2001	155	16,595	16,582	15,352	92.6
2002	153	16,785	16,744	15,180	90.7
2003	149	16,757	16,748	15,508	92.6
2004	149	16,894	16,974	15,783	93.0
2005	148	17,125	17,196	15,578	90.6
2006	149	17,339	17,385	15,602	89.7
2007	149	17,443	17,450	15,450	88.5
2008	150	17,594	17,607	15,027	85.3
2009	150	17,672	17,678	14,659	82.9
2010	148	17,584	[R]17,575	[R]15,177	[R]86.4
2011[P]	148	17,736	17,726	15,283	86.2

[a]Through 1956, includes only those refineries in operation on January 1; beginning in 1957, includes all "operable" refineries on January 1.
[b]Average of monthly capacity data.
[c]Through 1980, utilization is calculated by dividing gross input to distillation units by one-half of the sum of the current year's January 1 capacity and the following year's January 1 capacity. Beginning in 1981, utilization is calculated by dividing gross input to distillation units by the annual average capacity.
R = Revised. P = Preliminary. NA = Not available.

SOURCE: "Table 5.9. Refinery Capacity and Utilization, Selected Years, 1949–2011," in *Annual Energy Review 2011*, U.S. Energy Information Administration, September 27, 2012, http://www.eia.gov/totalenergy/data/annual/pdf/aer.pdf (accessed October 3, 2012)

of an economic recession) to a high of 95.6% in 1998. Even though capacity has fallen since 1998, in 2011 it was still high at 86.2%.

As Table 2.1 shows, fewer refineries were operating in the United States in the early 21st century than in the past, but the gross input to their distillation units increased. In 2011 the gross input was 15.3 mbpd, near the record high of 15.8 mbpd that was set in 2004. The number of refineries has dropped for a variety of reasons, including retirement of older inefficient plants and consolidation within the oil industry. In addition, some countries that export oil to the United States have begun refining their own oil and selling the petroleum products directly.

The last large refinery built in the United States was completed in 1976, and the last new refinery of any size began operation in Valdez, Alaska, in 1993.

As of December 2012, only one new refinery was undergoing an approval process. However, its future remained uncertain in the midst of lowered oil demand and public opposition. The Hyperion Energy Center was planning to locate the new refinery in South Dakota. In "Environmental Questions and Answers Regarding the Proposed Hyperion Energy Center" (February 25, 2011, http://denr.sd.gov/Hyperion/Air/20110225QAHyperion.pdf), the South Dakota Department of Environment and Natural Resources indicates that in 2010 the state's Board of Minerals and Environment approved Hyperion's request for an air permit to run the refinery; other permits, however, were still needed. According to Ben Dunsmoor, in "Hyperion: Committed to South Dakota Refinery" (October 3, 2012, http://www.keloland.com/newsdetail.cfm/hyperion-commit ted-to-south-dakota-refinery-/?id=138018), as of October 2012, a court challenge against the air permit was being heard by South Dakota's supreme court. In addition, the land deals that Hyperion made with local landowners at the site of the proposed refinery had expired, making it unlikely that the company would be able to continue with its original plan. Nevertheless, Dunsmoor notes that a company executive had stated publicly that Hyperion was still committed to constructing the oil refinery in South Dakota at an estimated price of around $10 billion.

DOMESTIC PRODUCTION
Petroleum Production

As noted earlier, the EIA broadly defines petroleum as including crude oil and other fossil fuel liquids. In *Annual Energy Review 2011*, the EIA indicates that total petroleum production in 1949 was 5.5 mbpd. This total included crude oil, lease condensate, natural gas plant liquids, and processing gain (refinery and blender net production minus refinery and blender net inputs). Since 1981 the EIA has also included biodiesel in its petroleum production statistics. (Biodiesel is a petroleum-like fuel produced from biological sources that are not fossil fuels.) Annual petroleum production between 1981 and 2011 peaked in 1985 at 11.2 mbpd. By 2005 it had plummeted to 8.1 mbpd, but rebounded to nearly 9.9 mbpd in 2011.

Crude Oil Production

The EIA reports in *Annual Energy Review 2011* that U.S. production of crude oil reached its highest level in 1970, at 9.6 mbpd. Over the next decade domestic production declined and leveled off at around 8.5 mbpd to 9 mbpd through 1985. (See Figure 2.7.) Production underwent a sustained decline from 1985 into the first decade of the 21st century. In 2008 production bottomed out at around 5 mbpd and then rose to 5.7 mbpd in 2011. Onshore production has historically accounted for the largest portion of total production. (See Figure 2.7.) In 2011 approximately 4.2 mbpd, or 74% of all U.S.-produced oil, was from onshore oil fields.

Figure 2.8 distinguishes production in Alaska from production in the 48 contiguous states. Note that an oil pipeline was constructed during the late 1970s to connect Alaska's oil-rich North Slope to the port of Valdez in the south of the state. In 1988 Alaskan production peaked at just over 2 mbpd, or 25% of the total 8.1 mbpd. In 2011 Alaskan production was 572,000 barrels per day (bpd), which was only about 10% of total production for that year. As explained in Chapter 7, certain pristine areas of northern Alaska are off limits to drilling. Unless these protected areas are opened for drilling, U.S. oil production there will likely continue to decline.

The decades-long decline in U.S. oil production shown in Figure 2.7 was due to a modest decline in the number of producing wells from 647,000 in 1985 to 497,000 in 2006. (See Figure 2.9.) More important was a steep drop in the number of barrels produced per day per well. According to the EIA, the average productivity in 1975 was 16.8 bpd per well. It reached a low point in 2008 of 9.4 bpd per well. By 2011 both the number of producing wells (536,000) and the average productivity (10.6 bpd per well) had increased somewhat.

Domestic oil production is affected by crude oil availability and legislative, legal, and cost issues related to drilling and extraction. For example, U.S. producers spend more money than do Middle Eastern producers to drill and extract crude oil because oil is available in enormous, easily accessible reservoirs in the Middle East. U.S. oil is not as easily recovered. When the cost of oil recovery severely reduces the profit margin on a barrel of oil, U.S. producers may shut down their most expensive wells, especially in times when oil prices are low. When oil prices are high, even the most expensive extraction methods, such as those used to recover oil from tar sands, are still cost effective.

FIGURE 2.7

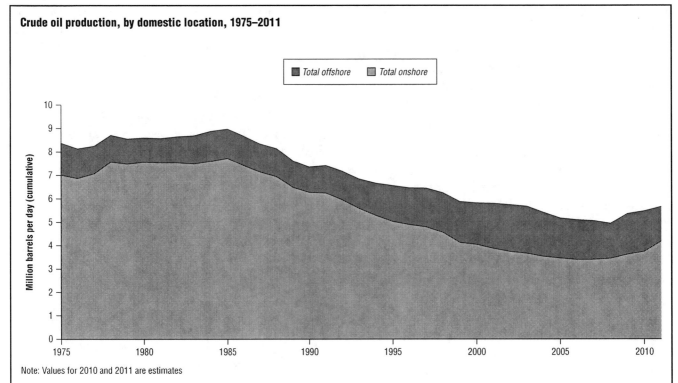

Crude oil production, by domestic location, 1975–2011

Note: Values for 2010 and 2011 are estimates

SOURCE: Adapted from "Table 5.2. Crude Oil Production and Crude Oil Well Productivity, Selected Years, 1954–2011," in *Annual Energy Review 2011*, U.S. Energy Information Administration, September 27, 2012, http://www.eia.gov/totalenergy/data/annual/pdf/aer.pdf (accessed October 3, 2012)

FIGURE 2.8

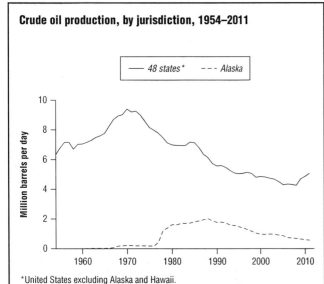

Crude oil production, by jurisdiction, 1954–2011

*United States excluding Alaska and Hawaii.
Notes: Crude oil includes lease condensate. The category called 48 states excludes Alaska and Hawaii and includes state onshore, state offshore, and federal offshore production. Alaska production include state onshore, state offshore, and federal offshore production.

SOURCE: "Figure 5.2. Crude Oil Production and Crude Oil Well Productivity, 1954–2011: Crude Oil Production, 48 States and Alaska," in *Annual Energy Review 2011*, U.S. Energy Information Administration, September 27, 2012, http://www.eia.gov/totalenergy/data/annual/pdf/aer.pdf (accessed October 3, 2012)

FIGURE 2.9

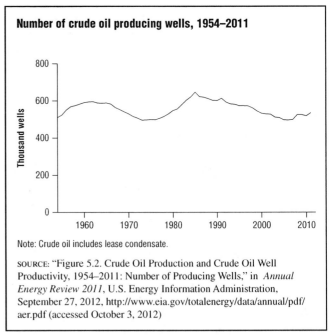

Number of crude oil producing wells, 1954–2011

Note: Crude oil includes lease condensate.

SOURCE: "Figure 5.2. Crude Oil Production and Crude Oil Well Productivity, 1954–2011: Number of Producing Wells," in *Annual Energy Review 2011*, U.S. Energy Information Administration, September 27, 2012, http://www.eia.gov/totalenergy/data/annual/pdf/aer.pdf (accessed October 3, 2012)

DOMESTIC CONSUMPTION

Domestic oil consumption is difficult to calculate with certainty. As noted earlier, crude oil is a raw material that is refined and processed into multiple petroleum products. In "Oil: Crude and Petroleum Products Explained" (September 24, 2012, http://www.eia.gov/energyexplained/index.cfm?page=oil_home), the EIA explains that a barrel

(42 gallons [159 L]) of crude oil provides approximately 45 gallons' (170 L) worth of petroleum products. The agency notes that "this gain from processing the crude oil is similar to what happens to popcorn, which gets bigger after it is popped."

The EIA notes in *Annual Energy Review 2011* that it uses petroleum product supply data to estimate petroleum consumption. These calculations are quite complex, and over time the agency has included more and more liquid fuels, such as biodiesel, in its estimates. Thus, consumption figures are not exactly comparable across years. The EIA estimates that domestic consumption in 2011 was 18.9 mbpd.

Figure 2.10 shows petroleum consumption by energy-using sector between 1973 and 2011. (Note that the sectors were defined in detail in Chapter 1.) Historically, the transportation sector has been, by far, the biggest consumer of petroleum, followed by the industrial sector, the residential and commercial sectors combined, and the electric power sector. The transportation sector consumed 13.2 mbpd in 2011, or 70% of the total consumption of 18.9 mbpd. (See Figure 2.11.) Consumption for industrial use was 4.5 mbpd (24% of the total), for residential use was 0.7 mbpd (4% of the total), for commercial use was 0.4 mbpd (2% of the total), and for electric power use was 0.1 mbpd (less than 1% of the total). Table 2.2 provides a breakdown by sector of petroleum product consumption for 2011.

FIGURE 2.10

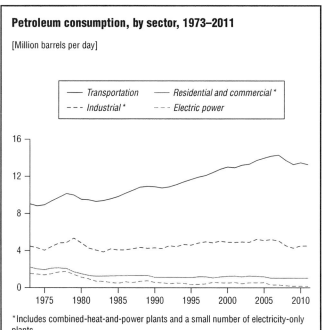

Petroleum consumption, by sector, 1973–2011

[Million barrels per day]

— Transportation   — Residential and commercial *
--- Industrial *   --- Electric power

*Includes combined-heat-and-power plants and a small number of electricity-only plants.

SOURCE: "Figure 3.7. Petroleum Consumption by Sector (Million Barrels per Day): By Sector, 1973–2011," in *Monthly Energy Review: September 2012*, U.S. Energy Information Administration, September 26, 2012, http://www.eia.gov/totalenergy/data/monthly/archive/00351209.pdf (accessed October 3, 2012)

Most petroleum used in the transportation sector is for motor gasoline. According to the EIA, motor gasoline has historically accounted for about 40% to 48% of total petroleum products supplied.

In 2011 overall petroleum consumption by product was as follows:

- Motor gasoline—8.7 mbpd (46% of total)
- Distillate fuel oil—3.8 mbpd (20% of total)
- Liquefied petroleum gases—2.2 mbpd (12% of total)
- Jet fuel—1.4 mbpd (7% of total)
- Residual fuel oil—0.5 mbpd (3% of total)
- Petroleum coke—0.4 mbpd (2% of total)
- Other—1.8 mbpd (10% of total)

WORLD OIL PRODUCTION AND CONSUMPTION

Table 2.3 shows the production of crude oil including lease condensate for 1980, 1990, 2000, 2010, and 2011 for the world, by region, and for the top-25 producers. The total world crude oil production increased from 59.6 mbpd in 1980 to 74.1 mbpd in 2011. Between 2000 and 2011 world production increased by 8%. The Middle East has been, by far, the largest producing region, accounting for 24 mbpd in 2011, or 32% of the world production. However, Eurasia experienced the highest production growth rate (62%) of any region between 2000 and 2011. Russia led all countries in production in 2011 with 9.8 mbpd, which was an increase of 51% since 2000. Saudi Arabia was second, and the United States was third. China and Iran completed the top-five producers.

The total world petroleum consumption is shown in Table 2.4. It grew from 63.1 mbpd in 1980 to 88.3 mbpd in 2011. Between 2000 and 2011 consumption grew by 15%. The Middle East experienced the highest growth in consumption between 2000 and 2011, with a 65% increase. However, the United States has been, by far, the leading petroleum consumer over the period shown. U.S. consumption of 18.8 mbpd in 2011 accounted for 21% of the total world consumption. Other top consumers in 2011 included China, Japan, India, and Saudi Arabia.

OIL IMPORTS AND EXPORTS

Countries that have surplus oil (e.g., Saudi Arabia) sell their excess to countries that need more than they can produce, such as the United States, China, Japan, and west European countries. They sell petroleum as both crude oil and refined products, although the trend has been moving toward refined products because they bring higher profits. The leading suppliers of petroleum to the United States in 2011 were Canada, Mexico, Saudi Arabia, Venezuela,

FIGURE 2.11

Petroleum flow, 2011

[Million barrels per day]

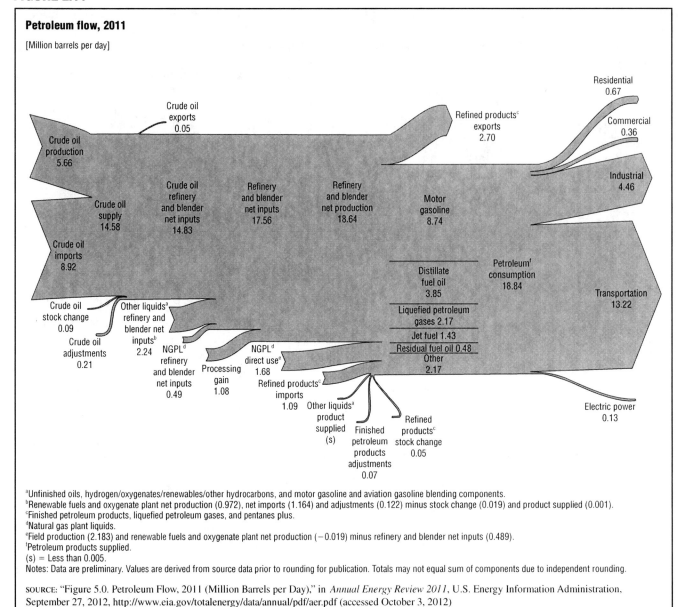

[a]Unfinished oils, hydrogen/oxygenates/renewables/other hydrocarbons, and motor gasoline and aviation gasoline blending components.
[b]Renewable fuels and oxygenate plant net production (0.972), net imports (1.164) and adjustments (0.122) minus stock change (0.019) and product supplied (0.001).
[c]Finished petroleum products, liquefied petroleum gases, and pentanes plus.
[d]Natural gas plant liquids.
[e]Field production (2.183) and renewable fuels and oxygenate plant net production (−0.019) minus refinery and blender net inputs (0.489).
[f]Petroleum products supplied.
(s) = Less than 0.005.
Notes: Data are preliminary. Values are derived from source data prior to rounding for publication. Totals may not equal sum of components due to independent rounding.

SOURCE: "Figure 5.0. Petroleum Flow, 2011 (Million Barrels per Day)," in *Annual Energy Review 2011*, U.S. Energy Information Administration, September 27, 2012, http://www.eia.gov/totalenergy/data/annual/pdf/aer.pdf (accessed October 3, 2012)

Nigeria, Russia, Iraq, Brazil, and the United Kingdom. (See Figure 2.12.)

Even though the United States produces a significant amount of petroleum, it has been importing oil since World War II (1939–1945). Initially, the imported amounts were very small. According to the EIA, in *Annual Energy Review 2011*, in 1949 net imports (imports minus exports) accounted for 5.5% of the estimated domestic consumption. This value rose steadily, peaking at 60.3% in 2005 before dropping to 44.7% in 2011.

At first, imported oil was cheap and readily available, suiting the demands of a growing American population and economy. Furthermore, relatively low world crude oil prices often resulted in reduced domestic oil production: when the world price was lower than the cost of

producing oil from some U.S. wells, domestic oil became unprofitable and was not produced. Consequently, more oil was imported.

The amount of oil imported to the United States between 1960 and 2011 is shown in Table 2.5. The table highlights imports from countries that are and are not members of the Organization of the Petroleum Exporting Countries (OPEC).

Organization of the Petroleum Exporting Countries

OPEC was formed in 1960 by five major oil-producing countries: Iran, Iraq, Kuwait, Saudi Arabia, and Venezuela. It is a cartel (a group of countries that agree to control production and marketing to avoid competing with one another). Since 1973 OPEC has tried to influence the worldwide oil supply to achieve higher prices. Over the

TABLE 2.2

Petroleum consumption, by sector and product, 2011

Transportation sector	
Motor gasoline	8,565
Distillate fuel oil	2,779
Jet fuel	1,425
Residual fuel oil	359
Lubricants	60
Liquefied petroleum gases	21
Aviation gasoline	15
Total	**13,224**
Industrial sector	
Liquefied petroleum gases	1,672
Other petroleum	1,300
Distillate fuel oil	568
Asphalt and road oil	355
Petroleum coke	307
Motor gasoline	143
Lubricants	64
Residual fuel oil	49
Kerosene	2
Total	**4,460**
Residential sector	
Liquefied petroleum gases	378
Distillate fuel oil	278
Kerosene	9
Total	**665**
Commercial sector	
Distillate fuel oil	196
Liquefied petroleum gases	100
Residual fuel oil	32
Motor gasoline	28
Kerosene	1
Petroleum coke	*
Total	**357**
Electric power sector	
Petroleum coke	60
Residual fuel oil	41
Distillate fuel oil	29
Total	**130**

Note: Petroleum production is used as an approximation of petroleum consumption. Values for 2011 are preliminary. Motor gasoline is finished motor gasoline. It also includes ethanol blended into motor gasoline.

SOURCE: Adapted from "Table 5.13a. Petroleum Consumption Estimates: Residential and Commercial Sectors, Selected Years, 1949–2011 (Thousand Barrels per Day)," "Table 5.13b. Petroleum Consumption Estimates: Industrial Sector, Selected Years, 1949–2011 (Thousand Barrels per Day)," "Table 5.13c. Petroleum Consumption Estimates: Transportation Sector, Selected Years, 1949–2011 (Thousand Barrels per Day)," and "Table 5.13d. Petroleum Consumption Estimates: Electric Power Sector, Selected Years, 1949–2011 (Thousand Barrels per Day)," in *Annual Energy Review 2011*, U.S. Energy Information Administration, September 27, 2012, http://www.eia.gov/totalenergy/data/annual/pdf/aer.pdf (accessed October 3, 2012)

decades various countries have joined and quit the cartel. As of 2012, OPEC had 12 member countries. (See Table 2.6.) Seven of the countries were located in the oil-rich Persian Gulf in the Middle East. All OPEC members except Libya and Ecuador were among the top-25 oil producers during 2011. (See Table 2.3.) Two OPEC members—Saudi Arabia and Iran—were the second- and fifth-largest producers, respectively. Saudi Arabia, in particular, has been a dominant oil producer for decades.

The United States was hugely dependent on OPEC oil during the 1960s and 1970s. (See Table 2.5.) In fact,

in 1977 just over 70% of all oil imported to the United States was from OPEC. During the early 1990s approximately half of the nation's oil was supplied by OPEC. By 2011 the percentage had declined to 39.9%.

CONCERN ABOUT FOREIGN OIL DEPENDENCY

During the 1970s U.S. leaders were concerned that so much of the country's economic structure, based heavily on imported oil, was dependent on decisions in OPEC countries. Oil resources became an issue of national security, and OPEC countries, especially the Arab members, were often portrayed as potentially strangling the U.S. economy. Efforts were made to reduce imports by raising public awareness and by encouraging industry to create more energy-efficient products, such as automobiles with better gas mileage. These measures are described in detail in Chapter 9. Nonetheless, total petroleum imports rose through the end of the 1970s. (See Table 2.5.) Imports declined and then rebounded over the following decades as conservation efforts waned in the face of low oil prices. In fact, fuel efficiency gains in automobiles during the 1990s were offset by the public's growing preference for large vehicles, such as sport-utility vehicles. As a result, total imports of petroleum grew steadily.

Concern that U.S. dependence on foreign oil, especially OPEC oil, represented a threat to national security or national stability grew after the terrorist attacks of September 11, 2001. That concern was heightened by the war on terror, the war with Iraq, and the consequent unrest in Middle Eastern nations. Regardless, demand for the product persisted, and total petroleum imports reached a peak of 13.7 mbpd in 2005 and 2006. (See Table 2.5.) By the end of 2011 that level had dropped to 11.4 mbpd.

STRATEGIC PETROLEUM RESERVE

Early in the 20th century the Naval Petroleum Reserve was established to ensure that the U.S. Navy would have adequate fuel in the event of war. Large tracts of government land with known deposits of oil were set aside. In 1975, in response to the growing concern over U.S. energy dependence, Congress expanded this concept by creating the Strategic Petroleum Reserve. Oil and refined products are stored in 41 deep salt caverns in Louisiana and Texas. (The caverns are used because oil does not dissolve salt the way water does.) If the United States suddenly finds its supplies cut off, the reserves can be connected to existing pipelines and the oil pumped out.

At the end of 2011 the Strategic Petroleum Reserve contained 696 million barrels of oil. (See Figure 2.13.) According to the EIA, in *Annual Energy Review 2011*, this store was equal to approximately 82 days' worth of imported oil. There has been a decline in the reserves in terms of days' worth of net imports, from a high of 115 days in 1985, to a low of 50 days in 2001, to 82 days in

TABLE 2.3

World production of crude oil, including lease condensate, by region and selected country, selected years, 1980–2011

[Thousand barrels per day]

	1980	1990	2000	2010	2011	% change between 2000 and 2011
World	**59,558**	**60,497**	**68,522**	**74,005**	**74,080**	**8%**
Middle East	18,442	16,545	21,823	22,746	24,035	10%
Eurasia	11,706	10,975	7,810	12,577	12,649	62%
North America	11,968	11,461	10,903	10,831	11,158	2%
Africa	6,125	6,437	7,487	9,874	8,554	14%
Asia & Oceania	4,848	6,468	7,538	7,819	7,631	1%
Central & South America	3,647	4,318	6,598	6,401	6,635	1%
Europe	2,821	4,292	6,364	3,757	3,418	−46%
Russia	—	—	6,479	9,694	9,774	51%
Saudi Arabia	9,900	6,410	8,404	8,900	9,458	13%
United States	8,597	7,355	5,822	5,479	5,658	−3%
China	2,114	2,774	3,249	4,078	4,069	25%
Iran	1,662	3,088	3,696	4,080	4,054	10%
Canada	1,435	1,553	1,977	2,732	2,904	47%
United Arab Emirates	1,709	2,117	2,368	2,415	2,688	14%
Iraq	2,514	2,040	2,571	2,399	2,626	2%
Mexico	1,936	2,553	3,104	2,621	2,596	−16%
Kuwait	1,656	1,175	2,079	2,300	2,530	22%
Nigeria	2,055	1,810	2,165	2,455	2,525	17%
Venezuela	2,168	2,137	3,155	2,146	2,240	−29%
Brazil	182	631	1,269	2,055	2,105	66%
Angola	150	475	746	1,939	1,786	139%
Norway	486	1,630	3,222	1,869	1,752	−46%
Kazakhstan	—	—	718	1,525	1,553	116%
Algeria	1,106	1,180	1,214	1,540	1,540	27%
Qatar	472	406	737	1,127	1,296	76%
United Kingdom (Offshore)	—	—	2,207	1,214	1,012	−54%
Azerbaijan	—	—	280	1,035	983	251%
Indonesia	1,577	1,462	1,428	953	918	−36%
Colombia	126	440	691	786	914	32%
Oman	282	685	970	865	886	−9%
India	182	660	646	751	782	21%
Argentina	491	483	761	632	588	−23%

SOURCE: Adapted from "Table. Production of Crude Oil including Lease Condensate (Thousand Barrels per Day)," in *International Energy Statistics*, U.S. Energy Information Administration, 2012, http://www.eia.gov/cfapps/ipdbproject/iedindex3.cfm?tid=5&pid=57&aid=1&cid=regions&syid=1980&eyid=2011&unit=TBPD (accessed October 7, 2012)

2011. This decline reflects the country's increasing reliance on imports since 1985. As the nation has imported a greater amount of oil, the days of net import replacement represented by the amount of oil in the reserves have dropped, even though the amount of oil in the reserves has increased.

OIL PRICES

When discussing oil prices, it is important to understand that there is no single price for a barrel of oil in the marketplace. Instead, analysts and investors monitor several benchmark prices. Table 2.7 shows various prices for crude oil and for liquid fuels produced from crude oil for the first quarter of 2011 through the second quarter of 2012. The West Texas Intermediate (WTI) spot average for crude oil was around $93 per barrel as of the second quarter 2012. WTI oil is a light sweet grade, and its price serves as a reference point for other U.S. oil prices. Another benchmark price of note is the Brent spot average. It represents a light sweet crude oil sourced from the North Sea in northern Europe. As of the second quarter of 2012 the Brent spot price for crude oil was

around $108 per barrel. The two other crude oil prices shown in Table 2.7 are the imported average (the average price of a barrel of crude oil imported to the United States) and the refiner average acquisition cost (the average price paid by U.S. refiners for a barrel of crude oil).

Figure 2.14 shows crude oil refiner acquisition costs between 1968 and 2011. According to the EIA, in *Annual Energy Review 2011*, the costs include the price of the oil purchased plus transportation costs and fees. The costs are expressed in nominal dollars (current to that year) and in real dollars (inflation adjusted). The real dollar amounts assume that the value of a dollar was constant over time. Thus, changes reflect actual market variations and not inflationary effects. Refiner acquisition costs hovered around $20 per barrel through the early 1970s and then began to grow. (See Figure 2.14.) There was a dramatic spike during the early 1980s to greater than $60 per barrel. However, by the mid-1980s the cost was back down to around $20 per barrel, a level maintained rather regularly through the end of the 1990s. The dawn of the 21st century witnessed a steep rise in refiner acquisition costs.

TABLE 2.4

World petroleum consumption, by region and selected country, selected years, 1980–2011

[Thousand barrels, per day]

	1980	1990	2000	2010	2011	% change between 2000 and 2011
World	63,120	66,525	76,791	87,231	88,347	15%
Asia & Oceania	10,729	13,818	20,827	27,029	28,828	38%
North America	20,208	20,319	23,819	23,480	23,161	−3%
Europe	16,056	14,695	15,913	15,317	15,033	−6%
Middle East	2,044	3,469	4,788	7,217	7,888	65%
Central & South America	3,613	3,761	5,213	6,261	6,412	23%
Eurasia	8,995	8,392	3,721	4,431	4,280	15%
Africa	1,474	2,071	2,499	3,401	3,505	40%
United States	17,056	16,988	19,701	19,180	18,835	−4%
China	1,765	2,296	4,796	9,392	9,850	105%
Japan	4,960	5,315	5,515	4,437	4,464	−19%
India	643	1,168	2,127	3,116	3,426	61%
Saudi Arabia	610	1,107	1,537	2,650	2,986	94%
Brazil	1,148	1,466	2,166	2,560	2,793	29%
Russia	—	—	2,578	3,038	2,725	6%
Germany	—	—	2,767	2,470	2,400	−13%
Canada	1,873	1,722	2,014	2,237	2,259	12%
Korea, South	537	1,048	2,135	2,268	2,230	4%
Mexico	1,270	1,599	2,096	2,080	2,133	2%
Iran	590	1,003	1,248	1,800	2,028	62%
France	2,256	1,826	2,000	1,831	1,792	−10%
United Kingdom	1,725	1,776	1,765	1,630	1,608	−9%
Italy	1,934	1,868	1,854	1,544	1,454	−22%
Spain	990	1,010	1,433	1,441	1,384	−3%
Indonesia	408	651	1,037	1,357	1,119	8%
Australia	594	738	872	989	1,023	17%
Netherlands	792	734	855	1,020	1,010	18%
Venezuela	400	396	500	765	980	96%
Thailand	224	407	725	960	942	30%
Taiwan	380	542	865	961	929	7%
Singapore	202	363	660	1,080	896	36%
Iraq	217	400	462	694	818	77%
Turkey	314	477	667	675	706	6%

SOURCE: Adapted from "Table. Total Petroleum Consumption (Thousand Barrels per Day)," in *International Energy Statistics*, U.S. Energy Information Administration, 2012, http://www.eia.gov/cfapps/ipdbproject/iedindex3.cfm?tid=5&pid=5&aid=2&cid=regions&syid=1980&eyid=2011&unit=TBPD (accessed October 7, 2012)

FIGURE 2.12

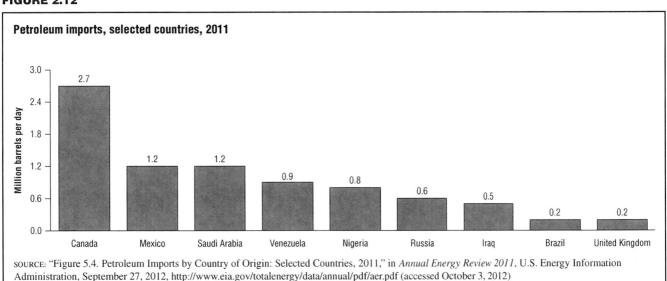

Petroleum imports, selected countries, 2011

SOURCE: "Figure 5.4. Petroleum Imports by Country of Origin: Selected Countries, 2011," in *Annual Energy Review 2011*, U.S. Energy Information Administration, September 27, 2012, http://www.eia.gov/totalenergy/data/annual/pdf/aer.pdf (accessed October 3, 2012)

Between 2000 and 2008 the cost soared from around $30 per barrel to nearly $100 per barrel. As noted in Chapter 1, the United States suffered a severe economic downturn called the Great Recession, which lasted from December 2007 to June 2009. Consumer demand dropped severely during this period. Correspondingly,

TABLE 2.5

Petroleum imports, selected countries and years, 1960–2011

Year	Persian Gulf[a]	Selected OPEC countries — Iraq	Nigeria	Saudi Arabia[b]	Venezuela	Total OPEC[c]	Selected non-OPEC countries — Brazil	Canada	Mexico	Russia[d]	United Kingdom	Total non-OPEC[c]	Total imports	Imports from Persian Gulf[a] as share of total imports	Imports from OPEC as share of total imports
	Thousand barrels per day													Percent	
1960	RE326	22	e	84	911	1,233	1	120	16	0	(s)	581	1,815	RE17.9	68.0
1965	R359	16	e	158	994	1,439	0	323	48	0	(s)	1,029	2,468	R14.5	58.3
1966	R319	26	e	147	1,018	1,444	0	384	45	0	6	1,129	2,573	R12.4	56.1
1967	R203	5	e	92	938	1,247	2	450	49	0	11	1,290	2,537	R8.0	49.2
1968	R218	0	e	74	886	1,287	(s)	506	45	0	28	1,553	2,840	R7.7	45.3
1969	R193	0	e	65	875	1,286	0	608	43	2	20	1,879	3,166	R6.1	40.6
1970	R184	0	e	30	989	1,294	2	766	42	3	11	2,126	3,419	R5.4	37.8
1971	R379	11	102	128	1,020	1,673	3	857	27	0	10	2,253	3,926	R9.7	42.6
1972	471	4	251	190	959	2,046	5	1,108	21	8	9	2,695	4,741	9.9	43.2
1973	848	4	459	486	1,135	2,993	9	1,325	16	26	15	3,263	6,256	13.6	47.8
1974	1,039	0	713	461	979	3,256	2	1,070	8	20	8	2,856	6,112	17.0	53.3
1975	1,165	2	762	715	702	3,601	5	846	71	14	14	2,454	6,056	19.2	59.5
1976	1,840	26	1,025	1,230	700	5,066	0	599	87	11	31	2,247	7,313	25.2	69.3
1977	2,448	74	1,143	1,380	690	6,193	0	517	179	12	126	2,614	8,807	27.8	70.3
1978	2,219	62	919	1,144	646	5,751	0	467	318	2	180	2,612	8,363	26.5	68.8
1979	2,069	88	1,080	1,356	690	5,637	1	538	439	1	202	2,819	8,456	24.5	66.7
1980	1,519	28	857	1,261	481	4,300	3	455	533	1	176	2,609	6,909	22.0	62.2
1981	1,219	(s)	620	1,129	406	3,323	23	447	522	5	375	2,672	5,996	20.3	55.4
1982	696	3	514	552	412	2,146	47	482	685	1	456	2,968	5,113	13.6	42.0
1983	442	10	302	337	422	1,862	41	547	826	1	382	3,189	5,051	8.8	36.9
1984	506	12	216	325	548	2,049	60	630	748	13	402	3,388	5,437	9.3	37.7
1985	311	46	293	168	605	1,830	61	770	816	8	310	3,237	5,067	6.1	36.1
1986	912	81	440	685	793	2,837	50	807	699	18	350	3,387	6,224	14.7	45.6
1987	1,077	83	535	751	804	3,060	84	848	655	11	352	3,617	6,678	16.1	45.8
1988	1,541	345	618	1,073	794	3,520	98	999	747	29	315	3,882	7,402	20.8	47.6
1989	1,861	449	815	1,224	873	4,140	82	931	767	48	215	3,921	8,061	23.1	51.4
1990	1,966	518	800	1,339	1,025	4,296	49	934	755	45	189	3,721	8,018	24.5	53.6
1991	1,845	0	703	1,802	1,035	4,092	22	1,033	807	29	138	3,535	7,627	24.2	53.7
1992	1,778	0	681	1,720	1,170	4,092	20	1,069	830	18	230	3,796	7,888	22.5	51.9
1993	1,782	0	740	1,414	1,300	4,273	33	1,181	919	55	350	4,347	8,620	20.7	49.6
1994	1,728	0	637	1,402	1,334	4,247	31	1,272	984	30	458	4,749	8,996	19.2	47.2
1995	1,573	0	627	1,344	1,480	4,002	8	1,332	1,068	25	383	4,833	8,835	17.8	45.3
1996	1,604	1	617	1,363	1,676	4,211	9	1,424	1,244	25	308	5,267	9,478	16.9	44.4
1997	1,755	89	698	1,407	1,773	4,569	5	1,563	1,385	13	226	5,593	10,162	17.3	45.0
1998	2,136	336	696	1,491	1,719	4,905	26	1,598	1,351	24	250	5,803	10,708	19.9	45.8
1999	2,464	725	657	1,478	1,493	4,953	26	1,539	1,324	89	365	5,899	10,852	22.7	45.6
2000	2,488	620	896	1,572	1,546	5,203	51	1,807	1,373	72	366	6,257	11,459	21.7	45.4
2001	2,761	795	885	1,662	1,553	5,528	82	1,828	1,440	90	324	6,343	11,871	23.3	46.6
2002	2,269	459	621	1,552	1,398	4,605	116	1,971	1,547	210	478	6,925	11,530	19.7	39.9
2003	2,501	481	867	1,774	1,376	5,162	108	2,072	1,623	254	440	7,103	12,264	20.4	42.1
2004	2,493	656	1,140	1,558	1,554	5,701	104	2,138	1,665	298	380	7,444	13,145	19.0	43.4
2005	2,334	531	1,166	1,537	1,529	5,587	156	2,181	1,662	410	396	8,127	13,714	17.0	40.7
2006	2,211	553	1,114	1,463	1,419	5,517	193	2,353	1,705	369	272	8,190	13,707	16.1	40.2
2007	2,163	484	1,134	1,485	1,361	5,980	200	2,455	1,532	414	277	7,489	13,468	16.1	44.4
2008	2,370	627	988	1,529	1,189	5,954	258	2,493	1,302	465	236	6,961	12,915	18.4	46.1
2009	1,689	450	809	1,004	1,063	4,776	309	2,479	1,210	563	245	6,915	11,691	14.4	40.9
2010	R1,711	R415	R1,023	1,096	R988	R4,906	R272	R2,535	R1,284	R612	256	R6,887	R11,793	14.5	41.6
2011P	1,862	460	817	1,195	944	4,534	249	2,706	1,205	621	158	6,825	11,360	16.4	39.9

TABLE 2.5

Petroleum imports, selected countries and years, 1960–2011 [CONTINUED]

[a]Bahrain, Iran, Iraq, Kuwait, Qatar, Saudi Arabia, United Arab Emirates, and the Neutral Zone (between Kuwait and Saudi Arabia).

[b]Through 1970, includes half the imports from the Neutral Zone. Beginning in 1971, includes imports from the Neutral Zone that are reported to U.S. Customs as originating in Saudi Arabia.

[c]On this table, "Total OPEC" for all years includes Iran, Iraq, Kuwait, Saudi Arabia, Venezuela, and the Neutral Zone (between Kuwait and Saudi Arabia); beginning in 1961, also includes Qatar; beginning in 1962, also includes Libya; for 1962–2008, also includes Indonesia; beginning in 1967, also includes United Arab Emirates; beginning in 1969, also includes Algeria; beginning in 1971, also includes Nigeria; for 1973–1992 and beginning in 2008, also includes Ecuador (although Ecuador rejoined OPEC in November 2007, on this table Ecuador is included in "Total Non-OPEC" for 2007); for 1975–1994, also includes Gabon; and beginning in 2007, also includes Angola. Data for all countries not included in "Total OPEC" reincluded in "Total Non-OPEC."

[d]Through 1992, may include imports from republics other than Russia in the former U.S.S.R.

[e]Nigeria joined OPEC in 1971. For 1960–1970, Nigeria is included in "Total Non-OPEC."

R = Revised. P = Preliminary. E = Estimate. (s) = Less than 500 barrels per day.

Notes: The country of origin for refined petroleum products may not be the country of origin for the crude oil from which the refined products were produced. For example, refined products imported from refineries in the Caribbean may have been produced from Middle East crude oil. Data include any imports for the Strategic Petroleum Reserve, which began in 1977. Totals may not equal sum of components due to independent rounding.

SOURCE: "Table 5.4. Petroleum Imports by Country of Origin, Selected Years, 1960–2011," in *Annual Energy Review 2011*, U.S. Energy Information Administration, September 27, 2012, http://www.eia.gov/totalenergy/data/annual/pdf/aer.pdf (accessed October 3, 2012)

TABLE 2.6

Organization of the Petroleum Exporting Countries (OPEC) and Persian Gulf nations, 2012

OPEC	Persian Gulf
• Iran	• Iran
• Iraq	• Iraq
• Kuwait	• Kuwait
• Saudi Arabia	• Saudi Arabia
• Qatar	• Qatar
• United Arab Emirates	• United Arab Emirates
• Algeria	• Bahrain
• Angola	
• Ecuador	
• Libya	
• Nigeria	
• Venezuela	

SOURCE: Adapted from "Did You Know?" in *Oil: Crude and Petroleum Products Explained—Oil Imports and Exports*, U.S. Energy Information Administration, May 1, 2012, http://www.eia.gov/energyexplained/index.cfm?page=oil_imports (accessed October 7, 2012).

FIGURE 2.13

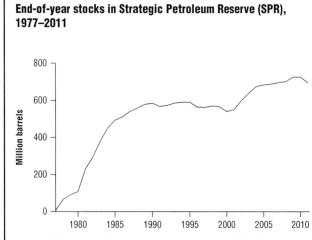

End-of-year stocks in Strategic Petroleum Reserve (SPR), 1977–2011

SOURCE: "Figure 5.17. Strategic Petroleum Reserve, 1977–2011: End-of-Year Stocks in SPR," in *Annual Energy Review 2011*, U.S. Energy Information Administration, September 27, 2012, http://www.eia.gov/totalenergy/data/annual/pdf/aer.pdf (accessed October 3, 2012).

refiner acquisition costs plummeted to less than $60 per barrel, but then immediately rebounded to greater than $90 per barrel.

Gasoline Prices

As noted earlier, motor gasoline has historically accounted for about 40% to 48% of total petroleum products supplied each year. Thus, the price of gasoline is very dependent on crude oil prices. In "Gasoline and Diesel Fuel Update" (December 3, 2012, http://www.eia.gov/petroleum/gasdiesel/), the EIA indicates the following average breakdown of costs for each gallon of regular gasoline that was sold to consumers in October 2012:

• Crude oil—64% of total

• Distribution and marketing—13% of total

• Refining—12% of total

• Taxes—11% of total

Table 2.8 shows the average annual gasoline prices for selected years between 1975 and 2011 and monthly prices for 2010, 2011, and January to August 2012. The prices are in nominal dollars, meaning they are not adjusted for inflation. The price of a gallon of unleaded regular gasoline more than doubled between 2000 and 2008 from $1.51 to $3.27. The average price in 2009 was lower at $2.35. By year-end 2011 it had crept back up to $3.53. As of August 2012, the price was $3.71 per gallon.

Oil Pricing Factors

As is true for all commodities, oil pricing is dependent on supply and demand factors. In general, when demand outpaces supply, prices go up. Likewise, when supply outpaces demand, prices go down. Higher prices encourage oil producers to produce more, such as by operating more expensive wells. However, higher prices tend to spur consumers to cut back on their petroleum consumption. This demand drop can then push prices lower. Energy markets are quite complex, and this is particularly true for oil because so much of it is imported into the United States. In addition, the U.S. government takes actions that affect the oil markets and hence oil pricing.

WEATHER CONDITIONS. The demand for petroleum products varies. Heating oil demand rises during the winter. A cold spell, which leads to a sharp rise in demand, may result in a corresponding price increase. A warm winter may be reflected in lower prices as suppliers try to clear out their inventory. Gasoline demand rises during the summer—people drive more for recreation— so gas prices rise as a consequence.

Weather disasters, such as Hurricane Katrina along the Gulf Coast during the summer of 2005, can disrupt production and refining capabilities and temporarily raise oil and petroleum product prices.

FOREIGN UNREST. Wars and other types of political unrest in oil-producing nations add volatility to petroleum prices, which fluctuate—sometimes dramatically— depending on the situation at the time. In "Major Disruptions of World Oil Supply" (September 3, 1998, http://www.eia.gov/emeu/25opec/sld001.htm), the EIA notes that six events occurred during the last half of the 20th century that caused major disruptions in the world's oil supply. One of the events was the Arab oil embargo, which is explained in the next section. The other five events were wars or other violent upheavals in the Middle East: the Suez War (1956), the Six Day War (1967), the Iran revolution (1979), the Iran-Iraq War (1980–1988), and the invasion of Kuwait by Iraq in 1990. Such events seriously affect the oil extraction and refining capabilities of the affected countries, which, in turn, affect oil prices.

TABLE 2.7

Prices for crude oil and petroleum products, 1st quarter 2011–2nd quarter 2012

	2011				2012	
	1st	2nd	3rd	4th	1st	2nd
Crude oil (dollars per barrel)						
West Texas intermediate spot average	93.50	102.22	89.72	93.99	102.88	93.42
Brent spot average	104.96	117.36	113.34	109.40	118.49	108.42
Imported average	94.23	108.74	102.06	105.36	108.13	101.33
Refiner average acquisition cost	94.01	108.13	100.61	104.55	107.62	101.53
Liquid fuels (cents per gallon)						
Refiner prices for resale						
Gasoline	267	312	297	271	297	299
Diesel fuel	286	316	307	304	317	301
Heating oil	275	305	295	296	312	293
Refiner prices to end users						
Jet fuel	287	322	308	302	321	304
No. 6 residual fuel oil (a)	217	246	249	250	270	266
Retail prices including taxes						
Gasoline regular grade (b)	330	380	363	337	361	372
Gasoline all grades (b)	335	385	369	342	367	378
On-highway diesel fuel	363	401	387	387	397	395
Heating oil	359	390	367	366	379	372

Prices are not adjusted for inflation.
(a) Average for all sulfur contents.
(b) Average self-service cash price.
Notes: Prices exclude taxes unless otherwise noted.
Historical data: Latest data available from Energy Information Administration databases supporting the following reports: *Petroleum Marketing Monthly*, DOE/EIA-0380; *Weekly Petroleum Status Report*, DOE/EIA-0208; *Natural Gas Monthly*, DOE/EIA-0130; *Electric Power Monthly*, DOE/EIA-0226; and *Monthly Energy Review*, DOE/EIA-0035. Natural gas Henry Hub and WTI crude oil spot prices from Reuter's News Service (http://www.reuters.com).
Minor discrepancies with published historical data are due to independent rounding.

SOURCE: Adapted from "Table 2. U.S. Energy Prices," in *Short Term Energy Outlook*, U.S. Energy Information Administration, September 11, 2012, http://www.eia.gov/forecasts/steo/pdf/steo_full.pdf (accessed October 7, 2012)

FIGURE 2.14

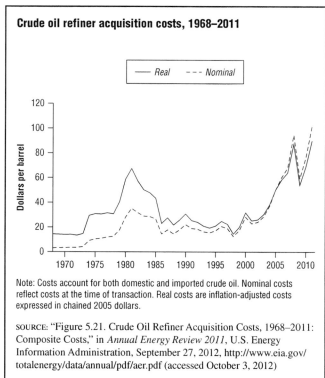

Crude oil refiner acquisition costs, 1968–2011

Note: Costs account for both domestic and imported crude oil. Nominal costs reflect costs at the time of transaction. Real costs are inflation-adjusted costs expressed in chained 2005 dollars.

SOURCE: "Figure 5.21. Crude Oil Refiner Acquisition Costs, 1968–2011: Composite Costs," in *Annual Energy Review 2011*, U.S. Energy Information Administration, September 27, 2012, http://www.eia.gov/totalenergy/data/annual/pdf/aer.pdf (accessed October 3, 2012)

FOREIGN PRICE MANIPULATION. Some top oil-producing countries have used their market power to manipulate oil prices upward for their own economic and political benefit. They do this by limiting the amount of oil they produce and/or export. Diminished supply (unless accompanied by diminished demand) increases the prices that buyers must pay.

The United States first fell victim to large-scale oil price manipulation in 1973, when some Arab countries, including those in OPEC, cut their oil exports to the United States in retaliation for U.S. aid to Israel during the Yom Kippur War, which was fought between Israel and neighboring Arab countries. The embargo lasted only six months, but the price of oil rose dramatically during that period. Americans experienced sudden price hikes for products that were produced from oil, such as gasoline and home heating oil, and faced temporary shortages. The energy problem quickly became an energy crisis, which led to occasional blackouts in cities and industries, temporary shutdowns of factories and schools, and frequent lines at gasoline service stations. The sudden increase in energy prices during the early 1970s is widely considered to have been a major cause of the economic recession of 1974 and 1975.

At other times, OPEC has used production quotas to put upward pressure on prices. However, this approach sometimes backfires. For example, high oil prices during the late 1970s to the mid-1980s encouraged conservation, which reduced demand for oil and led to a sharp decline

TABLE 2.8

Motor gasoline retail prices, U.S. city averages, 1975–2009 annual and January 2010–August 2012 monthly

[Dollars[a] per gallon, including taxes]

	Leaded regular	Unleaded regular	Unleaded premium[b]	All types[c]
1975 average	0.567	NA	NA	NA
1980 average	1.191	1.245	NA	1.221
1985 average	1.115	1.202	1.340	1.196
1990 average	1.149	1.164	1.349	1.217
1995 average	NA	1.147	1.336	1.205
1996 average	NA	1.231	1.413	1.288
1997 average	NA	1.234	1.416	1.291
1998 average	NA	1.059	1.250	1.115
1999 average	NA	1.165	1.357	1.221
2000 average	NA	1.510	1.693	1.563
2001 average	NA	1.461	1.657	1.531
2002 average	NA	1.358	1.556	1.441
2003 average	NA	1.591	1.777	1.638
2004 average	NA	1.880	2.068	1.923
2005 average	NA	2.295	2.491	2.338
2006 average	NA	2.589	2.805	2.635
2007 average	NA	2.801	3.033	2.849
2008 average	NA	3.266	3.519	3.317
2009 average	NA	2.350	2.607	2.401
2010				
January	NA	2.731	2.987	2.779
February	NA	2.659	2.922	2.709
March	NA	2.780	3.035	2.829
April	NA	2.858	3.113	2.906
May	NA	2.869	3.124	2.915
June	NA	2.736	3.000	2.783
July	NA	2.736	2.997	2.783
August	NA	2.745	3.015	2.795
September	NA	2.704	2.968	2.754
October	NA	2.795	3.055	2.843
November	NA	2.852	3.109	2.899
December	NA	2.985	3.234	3.031
Average	**NA**	**2.788**	**3.047**	**2.836**
2011				
January	NA	3.091	3.345	3.139
February	NA	3.167	3.424	3.215
March	NA	3.546	3.807	3.594
April	NA	3.816	4.074	3.863
May	NA	3.933	4.192	3.982
June	NA	3.702	3.972	3.753
July	NA	3.654	3.915	3.703
August	NA	3.630	3.893	3.680
September	NA	3.612	3.887	3.664
October	NA	3.468	3.745	3.521
November	NA	3.423	3.700	3.475
December	NA	3.278	3.553	3.329
Average	**NA**	**3.527**	**3.792**	**3.577**
2012				
January	NA	3.399	3.663	3.447
February	NA	3.572	3.840	3.622
March	NA	3.868	4.138	3.918
April	NA	3.927	4.194	3.976
May	NA	3.792	4.062	3.839
June	NA	3.552	3.825	3.602
July	NA	3.451	3.726	3.502
August	NA	3.707	3.991	3.759

TABLE 2.8

Motor gasoline retail prices, U.S. city averages, 1975–2009 annual and January 2010–August 2012 monthly [CONTINUED]

[Dollars[a] per gallon, including taxes]

[a]Prices are not adjusted for inflation.
[b]The 1981 average is based on September through December data only.
[c]Also includes types of motor gasoline not shown separately.
NA = Not available.
Notes: In September 1981, the Bureau of Labor Statistics changed the weights used in the calculation of average motor gasoline prices. From September 1981 forward, gasohol is included in the average for all types, and unleaded premium is weighted more heavily. Geographic coverage for 1973–1977 is 56 urban areas. Geographic coverage for 1978 forward is 85 urban areas.

SOURCE: Adapted from "Table 9.4. Motor Gasoline Retail Prices, U.S. City Average (Dollars per Gallon, Including Taxes)," in *Monthly Energy Review: September 2012*, U.S. Energy Information Administration, September 26, 2012, http://www.eia.gov/totalenergy/data/monthly/archive/00351209.pdf (accessed October 3, 2012)

petroleum market and the U.S. market. The United States obtained 39.9% of its imported oil from OPEC in 2011. (See Table 2.5.)

INVESTOR SPECULATION. Oil prices are also affected by investor speculation (trading that has high risks, but can reap large rewards). Oil is traded on commodities markets, such as the New York Mercantile Exchange, that offer futures contracts. These are basically agreements to buy or sell commodities on a set future date at a set price. Investors can profit from such transactions by gambling that price changes will occur in their favor before a futures contract expires.

In "Speculation in Crude Oil Adds $23.39 to the Price per Barrel" (*Forbes*, February 27, 2012), Robert Lenzner estimates that intense investor speculation in oil futures in February 2012 added approximately $23 to the price of a barrel of oil. He notes that the speculation was driven by concerns about "strife in Iran," which is a major oil producer. At that time, relations were extremely tense between Iran and the United States and its ally Israel over Iran's alleged development of nuclear weapons. In fact, there were some indications that Israel might attack Iran's nuclear facilities. Such an event would certainly reduce Iran's oil production, which would cut worldwide supply and hence raise oil prices. Thus, speculators anticipating this event were actually driving up the price of crude oil in February 2012 through their futures contracts. This example shows the extreme sensitivity of oil prices to international affairs. Even the hint of a possible problem arising in a major oil-producing country can cause oil prices to skyrocket.

GOVERNMENT INTERVENTION. U.S. government actions also affect oil pricing. Taxes push oil prices upward. For example, as noted earlier taxes accounted on average for 11% of the cost of each gallon of gasoline sold

in oil prices. As a result of the decreased demand for oil and lower prices, OPEC lost some of its ability to control its members and, consequently, prices. In addition, OPEC has faced increased competition since the 1970s from nonmember nations, such as Canada, Mexico, and Russia, that have expanded their oil production activities. Nevertheless, OPEC actions can still influence the worldwide

in October 2012. Other government actions push oil prices downward, mainly financial incentives, such as tax breaks and subsidies, that encourage domestic oil production.

The EIA indicates in *Direct Federal Financial Interventions and Subsidies in Energy in Fiscal Year 2010* (July 2011, http://www.eia.gov/analysis/requests/subsidy/pdf/subsidy.pdf) that the petroleum liquids and natural gas industry received $2.8 billion in direct expenditures, tax breaks, research and development funds, and other means of financial support from the federal government in 2010. This was around 8% of the $37.2 billion that was allocated to the energy industry as a whole that year.

The oil industry also benefits from liability limits on oil spills. Quinn Bowman explains in "Oil Spill Liability a Complicated Legal Web" (June 7, 2010, http://www.pbs.org/newshour/updates/politics/jan-june10/oillaw_06-04.html?print) that the Oil Pollution Act, as amended in 1990, limits to $75 million the damages that a responsible party must pay to "make up for lost economic activity, lost tax revenue and damage to natural resources as a result of an oil spill" from an offshore facility. (Note that the costs of "cleaning up and containing" spilled oil are separate costs and have no limit.) However, Bowman indicates that the $75 million damages limit does not apply if the responsible party is found to have committed "gross negligence, willful misconduct or a failure to comply with operating or safety regulations."

In addition, federal law limits liability amounts for tankers that spill oil in U.S. waters. In "Limits of Liability" (February 22, 2012, http://www.uscg.mil/npfc/Response/RPs/limits_of_liability.asp), the U.S. Coast Guard notes that the limits vary based on the type and tonnage of the vessels involved.

Critics suggest that all the oil spill liability limits are too low given the enormous economic and ecological consequences that a large oil spill can cause. Other provisions, however, are in place to help pay for damages from an oil spill. One is the Oil Spill Liability Trust Fund, which was created by Congress in 1986. According to the U.S. Coast Guard, in "The Oil Spill Liability Trust Fund (OSLTF)" (February 22, 2012, http://www.uscg.mil/npfc/About_NPFC/osltf.asp), the fund contained more than $1 billion in February 2012. Over the decades oil industry companies have sometimes been required to pay (depending on changing federal laws) into the fund via taxes on oil products. As of 2012, the tax was $0.08 per barrel of oil produced domestically or imported into the United States. The rate was scheduled to stay at this level until 2017, when it will increase to $0.09 per barrel.

ENVIRONMENTAL ISSUES

Environmental concerns related to the oil industry primarily involve the extraction, transport, refining, and combustion stages. Extraction impacts the environment via invasive drilling and mining to find and remove oil deposits. Crude oil is transported daily across the world's oceans and through a near-global network of pipelines. Even though large spills during drilling and transport are relatively rare, when they do occur they have devastating environmental and economic effects. Oil refining and the combustion of petroleum products have air quality impacts due to the release of contaminants such as carbon monoxide, nitrogen dioxide, sulfur dioxide, volatile organic compounds, and particulate matter. In addition, the combustion of oil, a carbon-rich fossil fuel, contributes to global warming and related climate change.

Oil Spills

In recent decades, the modern oil industry has experienced some environmental failures with serious consequences. Two of the most well-known events are the *Exxon Valdez* spill of 1989 and the British Petroleum (BP) spill in the Gulf of Mexico in 2010.

On March 24, 1989, the oil tanker *Exxon Valdez* hit a reef in Alaska and spilled 11 million gallons (41.6 million L) of crude oil into the waters of Prince William Sound. The spill was an environmental disaster for a formerly pristine area. Exxon was originally levied a $5 billion fine for damages, but that amount was later reduced by the U.S. Supreme Court to about $500 million.

The BP spill happened after an offshore oil platform located 40 miles (64 km) from the Louisiana coast exploded on April 20, 2010, and killed 11 workers. Engineers were unable to quickly cap the well on the ocean floor, which was about 1 mile (1.6 km) under the water's surface. Over a period of nearly three months millions of barrels of crude oil gushed into the Gulf, damaging the aquatic ecosystem and harming the area's fisheries and tourism industries. In November 2012 the U.S. government imposed a $4 billion fine on BP for damages. In addition, two BP oil rig supervisors were charged with manslaughter. The company and its business partners in the oil rig venture still faced private court cases and potentially billions of dollars more in damages. As of December 2012, the spill was considered to be the worst oil spill in U.S. history.

CHAPTER 3
NATURAL GAS

In 1821 William A. Hart drilled a well down 27 feet (8 m) and found a reservoir of natural gas in Fredonia, New York. This was the first successful modern natural gas well. Natural gas soon became popular as a fuel source for city street lighting, but by the turn of the 20th century it had been largely supplanted in this capacity by electricity. Following World War II (1939–1945), the United States constructed a massive pipeline network to carry natural gas to individual homes and businesses. The fuel is available in large amounts domestically at low prices and is widely used in industrial applications and for producing electricity.

UNDERSTANDING NATURAL GAS

The word *natural gas* is a generic term for a gaseous mixture of hydrocarbon compounds, mainly methane. This mixture is found naturally in certain geological formations and can be manufactured in relatively small quantities.

Naturally derived natural gas is divided into two categories, depending on its means of formation. Thermogenic natural gas results from the actions of pressure and heat over millions of years on prehistoric aquatic microorganisms. Thus, it is a fossil fuel. Biogenic natural gas results from the decomposition of organic (carbon-containing) matter by bacteria. This process can happen relatively quickly and is not limited to underground spaces; in fact, biogenic natural gas forms in landfills that contain organic garbage. When found underground, biogenic natural gas is typically at much shallower depths than is thermogenic natural gas.

Thermogenic Natural Gas

Thermogenic natural gas can be further categorized based on its geological source as conventional gas, coalbed methane, shale gas, or tight sand gas. (See Figure 3.1.)

CONVENTIONAL NATURAL GAS. Conventional natural gas, like crude oil, formed from prehistoric microscopic organisms that were converted to hydrocarbons deep underground by millennia of intense pressures and high temperatures. The gas was baked and squeezed more intensely than the crude oil. (See Figure 2.1 in Chapter 2.) Thus, the hydrocarbons in natural gas are smaller and lighter than those found in crude oil and are in a gaseous state. Methane, ethane, and propane are the primary constituents of natural gas, with methane making up the vast majority of the total. Once formed, crude oil and natural gas migrated away from their source rocks and toward the surface until they were either trapped beneath a nonporous rock formation (the seal shown in Figure 3.1) or seeped out into the open air. Conventional natural gas found trapped above oil pools is called associated gas, while conventional natural gas found by itself is called nonassociated gas.

COALBED METHANE. Coalbed methane is methane gas that naturally occurs as coal is formed. Coal is a dense solid fossil fuel that contains heavy hydrocarbons. In *Annual Energy Review 2011* (September 2012, http://www.eia.gov/totalenergy/data/annual/pdf/aer.pdf), the Energy Information Administration (EIA) within the U.S. Department of Energy notes that coalbed methane is trapped within the coal microstructure, but can be liberated and brought to the surface.

SHALE GAS. Shale is a dark dense type of sedimentary rock that is made up of fine tightly packed grains. Shale that contains high concentrations of organic matter is sometimes called black shale. The EIA notes in "What Is Shale Gas and Why Is It Important" (July 9, 2012, http://www.eia.gov/energy_in_brief/about_shale_gas.cfm) that shale gas is found in formations called plays, which are defined as "shale formations containing significant accumulations of natural gas and which share similar geologic and geographic properties." The natural gas

FIGURE 3.1

Schematic geology of natural gas resources

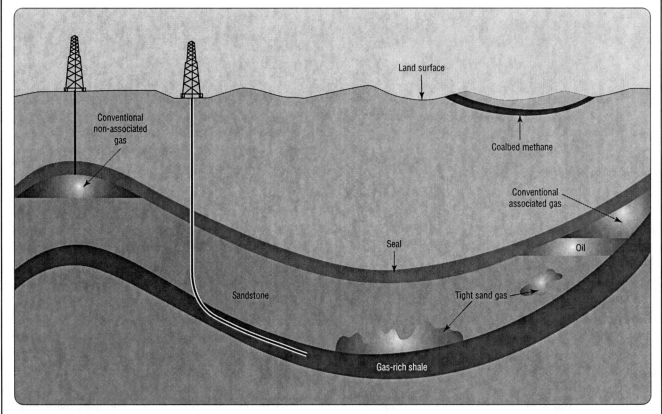

SOURCE: "Schematic Geology of Natural Gas Resources," in *Natural Gas Explained: Where Our Natural Gas Comes From*, U.S. Energy Information Administration, June 26, 2012, http://www.eia.gov/energyexplained/images/charts/NatGasSchematic-large.jpg (accessed October 7, 2012)

originally formed within the shale and became trapped in its pores. The gas is typically liberated by a process called fracking, which will be described later in this chapter.

TIGHT SAND GAS. In *Annual Energy Review 2011*, the EIA defines tight sand gas as "natural gas produced from a non-shale formation with extremely low permeability." Tight sand gas is found in sandstone, a type of sedimentary rock that is made up of sand-sized particles. (See Figure 3.1.) Sandstone is more porous than shale, but still typically requires fracking to release the natural gas that is trapped within it. Tight sand gas, or tight gas, is believed to have originated from the same thermogenic processes that produced conventional gas.

Natural Gas Deposits

Like crude oil deposits, natural gas deposits are found only in certain geological formations. Regions and countries that are known to have large natural gas deposits include North America, Russia, the Middle East, and northwestern Europe. Many of the earth's deposits are offshore (i.e., they lie beneath the oceans). Figure 2.3 and Figure 2.4 in Chapter 2 show the extent of offshore underground lands to which the United States claims

exclusive rights to extract resources, such as crude oil and natural gas. Many U.S. natural gas deposits lie in the nation's outer continental shelf, the outermost reaches of the continental shelf that are under federal control.

Estimates of the amounts of natural gas deposits that have not yet been extracted are provided in Chapter 7, along with information about natural gas exploration and development activities.

Measuring Natural Gas

In the United States natural gas volumes are measured in cubic feet. Because a cubic foot is a relatively small unit of measure, natural gas data are often presented in units of billion cubic feet (bcf) or trillion cubic feet (tcf). In countries that rely on the metric system, the cubic meter is the preferred unit for measuring natural gas volumes. Thus, natural gas data are typically expressed in units of billion cubic meters (bcm) or trillion cubic meters (tcm).

NATURAL GAS EXTRACTION

In many cases natural gas is extracted like crude oil, that is, via deeply drilled wells. Figure 3.2 shows the

FIGURE 3.2

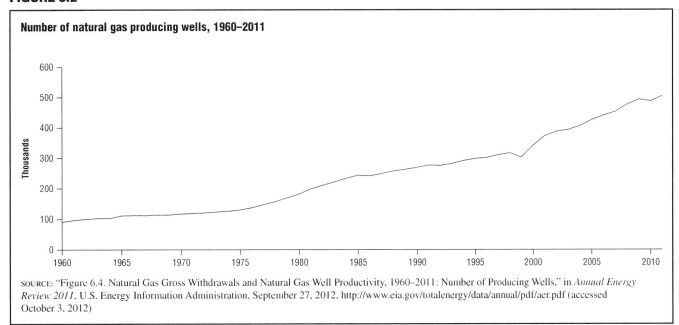

Number of natural gas producing wells, 1960–2011

SOURCE: "Figure 6.4. Natural Gas Gross Withdrawals and Natural Gas Well Productivity, 1960–2011: Number of Producing Wells," in *Annual Energy Review 2011*, U.S. Energy Information Administration, September 27, 2012, http://www.eia.gov/totalenergy/data/annual/pdf/aer.pdf (accessed October 3, 2012)

number of natural gas producing wells between 1960 and 2011. Even though the number of producing wells increased steadily after 1960 and more sharply after the mid-1970s, the number of gas wells in operation fluctuates from year to year because new wells are opened and old wells are closed. Weather and economic conditions also affect well operations. As of year-end 2011, the United States had an estimated 504,000 producing wells.

Figure 3.3 shows that some wells are dedicated solely to gas deposits, whereas others access both oil and gas deposits. (As noted in Chapter 2, crude oil and thermogenic natural gas are often found together.) The EIA estimates in *Annual Energy Review 2011* that nearly 80% of natural gas well withdrawal in 2011 was via wells that were dedicated to natural gas extraction. This included conventional wells, coalbed methane wells, and shale gas wells.

Figure 3.4 shows the average productivity of natural gas wells between 1960 and 2011. Productivity peaked during the early 1970s and then dropped through the mid-1980s. Ever since, it has remained at a relatively steady level, declining somewhat after 1999. At the peak of productivity in 1971, U.S. natural gas wells averaged 434,800 cubic feet (12,300 cubic meters) of natural gas per day per well. In 2011 they averaged 121,500 cubic feet (3,400 cubic meters) per day per well.

Fracking

Hydraulic fracturing (fracking) is a specialized well-based technique for extracting gas from unconventional geological sources (shale, tight sands, and coalbeds).

Figure 3.5 shows a typical fracking operation. A vertical well is drilled down and into the desired geological

formation. Then, a horizontal leg extends through the formation. Perforated pipes are inserted into the horizontal leg, and water is forced through the perforations under very high pressure, which causes numerous small fractures in the rock. The water contains sand, tiny ceramic beads, or other solids that serve as propping agents (they prop open the fractures so they do not close when the pressure is released and the water is removed). Natural gas then escapes from the rock fractures into the well and flows to the surface.

Fracking is a relatively new extraction process, in that it has only been used for several decades. It is controversial because large amounts of water are required and there is a danger that natural gas can be accidentally introduced into water aquifers.

PROCESSING AND TRANSPORTING NATURAL GAS

Most natural gas that is extracted from the ground is considered "wet" because it contains water vapor and hydrocarbons that easily liquefy at the cooler temperatures found above ground. The hydrocarbon liquids include lease condensate and natural gas plant liquids. Lease condensate is a liquid mix of heavy hydrocarbons that are recovered during natural gas processing at a lease, or field separation, facility. Natural gas plant liquids are a mixture of compounds such as propane and butane that are recovered as liquids at facilities later in the processing stage.

Consumer-grade natural gas is "dry," which means that it has been processed to remove water vapor, non-hydrocarbon gases (such as helium and nitrogen), and lease condensate and natural gas plant liquids. (See Figure 3.3.) Once processed, the dry gas flows into a

FIGURE 3.3

FIGURE 3.3

The natural gas industry

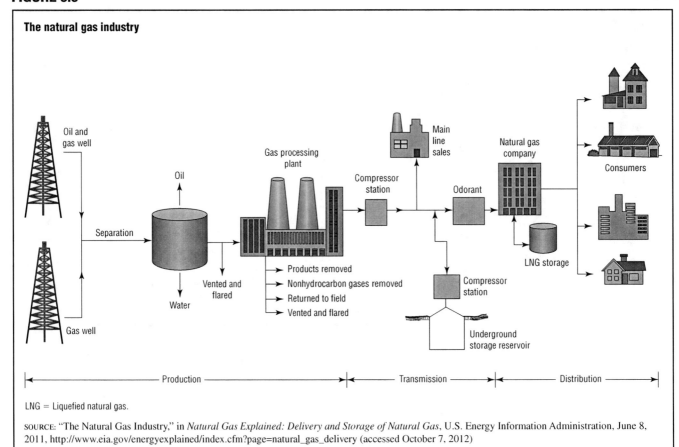

LNG = Liquefied natural gas.

SOURCE: "The Natural Gas Industry," in *Natural Gas Explained: Delivery and Storage of Natural Gas*, U.S. Energy Information Administration, June 8, 2011, http://www.eia.gov/energyexplained/index.cfm?page=natural_gas_delivery (accessed October 7, 2012)

FIGURE 3.4

Natural gas well average productivity, 1960–2011

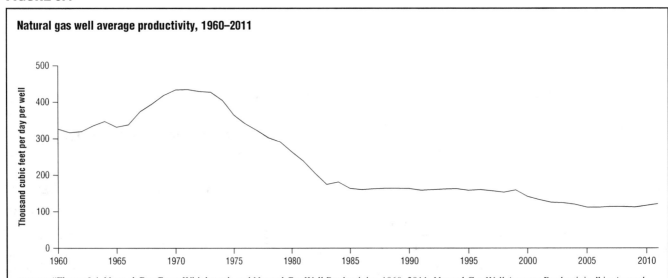

SOURCE: "Figure 6.4. Natural Gas Gross Withdrawals and Natural Gas Well Productivity, 1960–2011: Natural Gas Well Average Productivity," in *Annual Energy Review 2011*, U.S. Energy Information Administration, September 27, 2012, http://www.eia.gov/totalenergy/data/annual/pdf/aer.pdf (accessed October 3, 2012)

compressor station. In "Natural Gas Compressor Stations on the Interstate Pipeline Network: Developments Since 1996" (November 2007, http://www.eia.gov/pub/oil_gas/natural_gas/analysis_publications/ngcompressor/ngcom pressor.pdf), the EIA notes "the purpose of a compressor station is to boost the pressure in a natural gas pipeline and move the natural gas further downstream." In actuality, there are more than a thousand compressor stations

FIGURE 3.5

Hydraulic fracturing

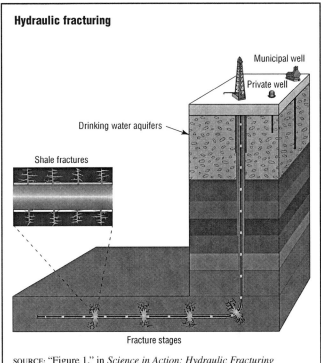

Fracture stages

SOURCE: "Figure 1," in *Science in Action: Hydraulic Fracturing Research Study*, U.S. Environmental Protection Agency, Office of Research and Development, June 2010, http://www.epa.gov/safewater/uic/pdfs/hfresearchstudyfs.pdf (accessed October 19, 2012)

situated along the nation's massive natural gas pipeline network. In addition, there are multiple underground storage reservoirs along the network.

Underground Storage

Because of seasonal, daily, and even hourly changes in demand, substantial natural gas storage facilities have been created. Many are depleted gas reservoirs that are located near transmission lines and marketing areas. Gas is injected into storage when market needs are lower than the available gas flow, and gas is withdrawn from storage when supplies from producing fields and the capacity of transmission lines are not adequate to meet peak demands. Figure 3.6 shows the amount of natural gas that was in underground storage between 1954 and 1958 and between 1962 and 2011. This natural gas consisted of base gas (permanently stored gas needed to maintain the proper pressure in the storage area) and working gas (gas that can be released from storage and used). The EIA indicates in *Annual Energy Review 2011* 7.8 tcf (219.9 bcm) was stored as of year-end 2011, including 4.3 tcf (121.9 bcm) of base gas and 3.5 tcf (98 bcm) of working gas. The United States had a total underground storage capacity of nearly 8.8 tcf (248.5 bcm) at year-end 2011.

Liquefied Natural Gas

Natural gas can be liquefied by cooling it to a very cold temperature. According to the EIA, in "Delivery and Storage

of Natural Gas" (June 8, 2011, http://www.eia.gov/energy explained/index.cfm?page=natural_gas_delivery), natural gas liquefies at around −260 degrees Fahrenheit (−162 degrees Celsius). Liquefied natural gas (LNG) can be transported or stored in tanks and returned to a gaseous state, as needed, by heating it above the liquefaction temperature. In "What Is LNG" (July 3, 2012, http://www.eia.gov/energyexplained/index.cfm?page=natural _gas_lng), the EIA explains that a liquefied volume of natural gas takes up far less space than the same volume of natural gas in a gaseous state. Liquefying is done so that the gas can be moved long distances, such as across the oceans, for importation and exportation. Of course, liquefaction is an energy-intensive process in that a lot of energy is required to cool the natural gas to the liquefaction temperature.

Transmission and Distribution

A vast network of natural gas pipelines crisscrosses the mainland United States. In "About U.S. Natural Gas Pipelines" (2012, http://www.eia.gov/pub/oil_gas/natural _gas/analysis_publications/ngpipeline/index.html), the EIA provides a map of this network. The agency also notes that the natural gas in this 305,000-mile (491,000-km) system generally flows northeastward, primarily from Texas and Louisiana, the two major gas-producing states, and from Oklahoma and New Mexico. It also flows west to California.

In its original state, natural gas is odorless. As such, an odorant is added to it before it is sold to consumers. (See Figure 3.3.) This odorant, which smells like rotten eggs, helps people to realize when gas is leaking around their homes or businesses so they can take action quickly by cutting off the gas flow. This safety precaution is necessary because natural gas is highly flammable and can cause tremendous explosions when ignited.

DOMESTIC WITHDRAWALS AND PRODUCTION

Figure 3.7 shows natural gas gross withdrawals and dry gas production between 1949 and 2011. Table 3.1 provides extraction and processing details on an annual basis. Gross withdrawals in 2011 were 28.6 tcf (809.2 bcm). Of this amount, 23 tcf (651.3 bcm) was dry gas produced for consumer use. Overall, dry gas production climbed dramatically from 1949 through the early 1970s and then fell through the mid-1980s. It increased at a moderate rate through the beginning of the 21st century and, after a brief decline, skyrocketed through 2011.

The vast majority of gross withdrawals between 1965 and 2011 were onshore. (See Figure 3.8.) During the mid-1970s and throughout much of the 1990s and into the early years of the first decade of the 21st century offshore withdrawals accounted for nearly a fourth of total withdrawals. However, since that time the offshore portion

FIGURE 3.6

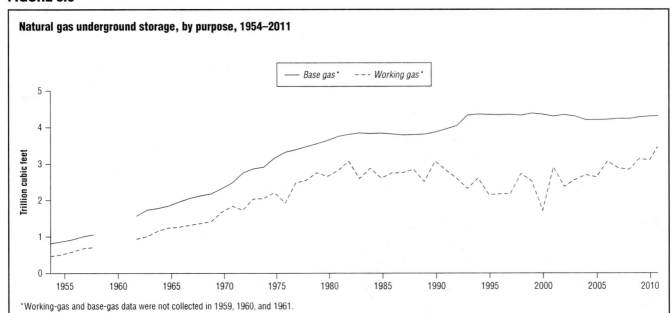

Natural gas underground storage, by purpose, 1954–2011

*Working-gas and base-gas data were not collected in 1959, 1960, and 1961.

SOURCE: "Figure 6.6. Natural Gas Underground Storage, End of Year: Base Gas and Working Gas in Underground Storage, 1954–2011," in *Annual Energy Review 2011*, U.S. Energy Information Administration, September 27, 2012, http://www.eia.gov/totalenergy/data/annual/pdf/aer.pdf (accessed October 3, 2012)

FIGURE 3.7

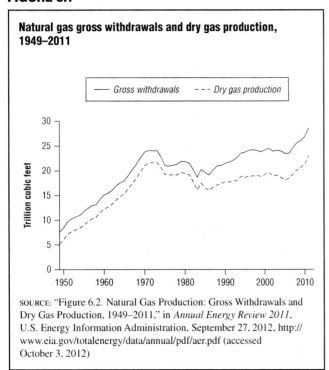

Natural gas gross withdrawals and dry gas production, 1949–2011

SOURCE: "Figure 6.2. Natural Gas Production: Gross Withdrawals and Dry Gas Production, 1949–2011," in *Annual Energy Review 2011*, U.S. Energy Information Administration, September 27, 2012, http://www.eia.gov/totalenergy/data/annual/pdf/aer.pdf (accessed October 3, 2012)

has been shrinking. In 2011 only about 8% of total withdrawals were from offshore locations.

Figure 3.9 categorizes gross withdrawals by location. Historically, Texas, Louisiana, and Oklahoma have been the three largest producing states. According to the EIA, in *Annual Energy Review 2011*, in 2011 Texas accounted for 7.9 bcf (224.6 million cubic meters) of gross

withdrawals, while Louisiana had 3.1 bcf (86.6 million cubic meters) and Oklahoma had 1.9 bcf (53.8 million cubic meters). The Federal Gulf of Mexico refers to the portion of the continental shelf that is subject to federal resource rights. As noted in Chapter 2, most coastal states have resource rights that extend outward 3 miles (5 km) from their coastlines. Resources beyond that limit fall under federal government control.

IMPORTS AND EXPORTS

Figure 3.10 compares domestic production and consumption between 1949 and 2011. The United States provides the vast majority of the natural gas consumed domestically. However, small amounts are imported and exported. In *Annual Energy Review 2011*, the EIA notes that net imports (imports minus exports) totaled only 1.9 tcf (55.1 bcm) in 2011.

Figure 3.11 shows natural gas imports and exports between 1949 and 2011. Imports of natural gas enter the United States via pipeline from Canada and Mexico. Liquefied natural gas arrives by tanker ships from various countries. The largest amount of natural gas that is imported to the United States comes from Canada. (See Figure 3.11.) According to the EIA, Canada supplied 3.1 tcf (87.8 bcm), or 90% of the total 3.5 tcf (97.8 bcm) that was imported in 2011. Likewise, Canada received 937 bcf (26.5 bcm), or 62% of the total 1.5 tcf (42.7 bcm) that was exported in 2011. The second-largest recipient of U.S. natural gas that year was Mexico, with 500 bcf (14.2 bcm), or 33% of the total.

TABLE 3.1

Natural gas production, by type of well, 1949–2011

[Billion cubic feet]

| Year | Natural gas gross withdrawals | | | | | Repressuring | Nonhydrocarbon gases removed | Vented and flared | Marketed production | Extraction loss[a] | Dry gas production |
	Natural gas wells	Crude oil wells	Coalbed wells	Shale gas wells	Total						
1949	4,986	2,561	NA	NA	7,547	1,273	NA	854	5,420	224	5,195
1950	5,603	2,876	NA	NA	8,480	1,397	NA	801	6,282	260	6,022
1955	7,842	3,878	NA	NA	11,720	1,541	NA	774	9,405	377	9,029
1960	10,853	4,234	NA	NA	15,088	1,754	NA	563	12,771	543	12,228
1965	13,524	4,440	NA	NA	17,963	1,604	NA	319	16,040	753	15,286
1970	18,595	5,192	NA	NA	23,786	1,376	NA	489	21,921	906	21,014
1975	17,380	3,723	NA	NA	21,104	861	NA	134	20,109	872	19,236
1976	17,191	3,753	NA	NA	20,944	859	NA	132	19,952	854	19,098
1977	17,416	3,681	NA	NA	21,097	935	NA	137	20,025	863	19,163
1978	17,394	3,915	NA	NA	21,309	1,181	NA	153	19,974	852	19,122
1979	18,034	3,849	NA	NA	21,883	1,245	NA	167	20,471	808	19,663
1980	17,573	4,297	NA	NA	21,870	1,365	199	125	20,180	777	19,403
1981	17,337	4,251	NA	NA	21,587	1,312	222	98	19,956	775	19,181
1982	15,809	4,463	NA	NA	20,272	1,388	208	93	18,582	762	17,820
1983	14,153	4,506	NA	NA	18,659	1,458	222	95	16,884	790	16,094
1984	15,513	4,754	NA	NA	20,267	1,630	224	108	18,304	838	17,466
1985	14,535	5,071	NA	NA	19,607	1,915	326	95	17,270	816	16,454
1986	14,154	4,977	NA	NA	19,131	1,838	337	98	16,859	800	16,059
1987	14,807	5,333	NA	NA	20,140	2,208	376	124	17,433	812	16,621
1988	15,467	5,532	NA	NA	20,999	2,478	460	143	17,918	816	17,103
1989	15,709	5,366	NA	NA	21,074	2,475	362	142	18,095	785	17,311
1990	16,054	5,469	NA	NA	21,523	2,489	289	150	18,594	784	17,810
1991	16,018	5,732	NA	NA	21,750	2,772	276	170	18,532	835	17,698
1992	16,165	5,967	NA	NA	22,132	2,973	280	168	18,712	872	17,840
1993	16,691	6,035	NA	NA	22,726	3,103	414	227	18,982	886	18,095
1994	17,351	6,230	NA	NA	23,581	3,231	412	228	19,710	889	18,821
1995	17,282	6,462	NA	NA	23,744	3,565	388	284	19,506	908	18,599
1996	17,737	6,376	NA	NA	24,114	3,511	518	272	19,812	958	18,854
1997	17,844	6,369	NA	NA	24,213	3,492	599	256	19,866	964	18,902
1998	17,729	6,380	NA	NA	24,108	3,427	617	103	19,961	938	19,024
1999	17,590	6,233	NA	NA	23,823	3,293	615	110	19,805	973	18,832
2000	17,726	6,448	NA	NA	24,174	3,380	505	91	20,198	1,016	19,182
2001	18,129	6,371	NA	NA	24,501	3,371	463	97	20,570	954	19,616
2002	17,795	6,146	NA	NA	23,941	3,455	502	99	19,885	957	18,928
2003	17,882	6,237	NA	NA	24,119	3,548	499	98	19,974	876	19,099
2004	17,885	6,084	NA	NA	23,970	3,702	654	96	19,517	927	18,591
2005	17,472	5,985	NA	NA	23,457	3,700	711	119	18,927	876	18,051
2006	17,996	5,539	NA	NA	23,535	3,265	731	129	19,410	906	18,504
2007	17,065	5,818	1,780	NA	24,664	3,663	661	143	20,196	930	19,266
2008	15,618	5,747	1,986	2,284	25,636	3,639	719	167	21,112	953	20,159
2009	R14,885	R5,812	1,977	3,384	R26,057	3,522	722	165	R21,648	1,024	P20,624
2010	b R20,841	R5,995	b	b	R26,836	R3,432	R837	R166	R22,402	R1,070	P21,332
2011	22,378	6,199	b	b	28,576	3,410	E831	E165	E24,170	P1,169	E23,000

[a]Volume reduction resulting from the removal of natural gas plant liquids, which are transferred to petroleum supply.
[b]Beginning in 2010, natural gas gross withdrawals from coalbed wells and shale gas wells are included in "Natural Gas Wells."
R = Revised. P = Preliminary. E = Estimate. NA = Not available.
Notes: Beginning with 1965 data, all volumes are shown on a pressure base of 14.73 p.s.i.a. at 60°F.
For prior years, the pressure base was 14.65 p.s.i.a. at 60°F. Totals may not equal sum of components due to independent rounding.

SOURCE: "Table 6.2. Natural Gas Production, Selected Years, 1949–2011 (Billion Cubic Feet)," in *Annual Energy Review 2011*, U.S. Energy Information Administration, September 27, 2012, http://www.eia.gov/totalenergy/data/annual/pdf/aer.pdf (accessed October 3, 2012)

FIGURE 3.8

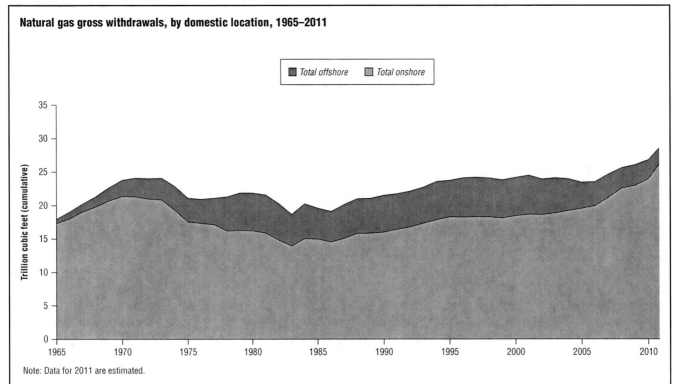

Natural gas gross withdrawals, by domestic location, 1965–2011

Note: Data for 2011 are estimated.

SOURCE: Adapted from "Table 6.4. Natural Gas Gross Withdrawals and Natural Gas Well Productivity, Selected Years, 1960–2011," in *Annual Energy Review 2011*, U.S. Energy Information Administration, September 27, 2012, http://www.eia.gov/totalenergy/data/annual/pdf/aer.pdf (accessed October 3, 2012)

FIGURE 3.9

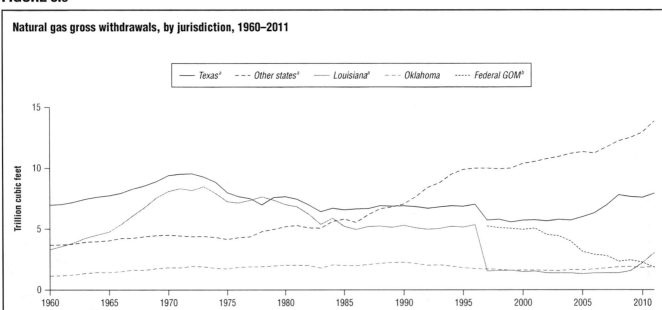

Natural gas gross withdrawals, by jurisdiction, 1960–2011

[a]Through 1996, includes gross withdrawals in Federal offshore areas of the Gulf of Mexico; beginning in 1997, these are included in "Federal Gulf of Mexico."
[b]Gulf of Mexico.

SOURCE: "Figure 6.4. Natural Gas Gross Withdrawals and Natural Gas Well Productivity, 1960–2011: Gross Withdrawals by State and Federal Gulf of Mexico," in *Annual Energy Review 2011*, U.S. Energy Information Administration, September 27, 2012, http://www.eia.gov/totalenergy/data/annual/pdf/aer.pdf (accessed October 3, 2012)

DOMESTIC CONSUMPTION

Natural gas fulfills an important part of the country's energy needs. It is an attractive fuel not only because its price is relatively low but also because it burns relatively cleanly and efficiently, which helps the country meet its environmental goals.

FIGURE 3.10

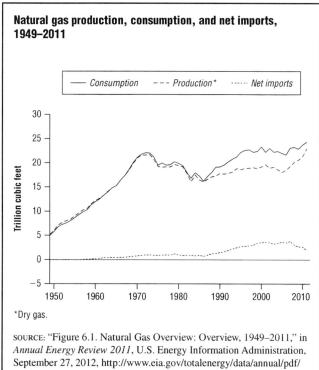

Natural gas production, consumption, and net imports, 1949–2011

*Dry gas.

SOURCE: "Figure 6.1. Natural Gas Overview: Overview, 1949–2011," in *Annual Energy Review 2011*, U.S. Energy Information Administration, September 27, 2012, http://www.eia.gov/totalenergy/data/annual/pdf/aer.pdf (accessed October 3, 2012)

Natural gas consumption rose between 1949 and 1972 and then generally declined through 1986. (See Figure 3.10.) Since 1986 natural gas consumption has been rising, hitting an all-time high of 24.4 tcf (690.1 bcm) in 2011.

Figure 3.12 shows natural gas consumption by energy-using sector between 1949 and 2011. The industrial sector has historically been the largest consumer; however, its dominance is being challenged by the electric power sector.

The EIA notes in "Uses of Natural Gas" (May 21, 2012, http://www.eia.gov/energyexplained/index.cfm?page=natural _gas_use) that the industrial sector uses natural gas to produce products that require large amounts of heat for manufacture. Examples include steel, glass, and paper. In addition, natural gas hydrocarbons are used as raw materials in products such as paints, plastics, and medicines. Consumption in the industrial sector was 8.2 tcf (230.8 bcm) in 2011. (See Figure 3.12 and Figure 3.13.) In general, natural gas usage by the industrial sector has remained relatively flat since the 1970s.

The electric power sector was the second-largest natural gas consumer in 2011 and has held that ranking since 1998, when it surpassed the residential sector. As will be explained in Chapter 8, the share of fossil-fired electric power generation held by natural gas grew tremendously between the 1980s and 2011 because of a variety of economic factors.

The residential and commercial sectors ranked third and fourth, respectively, in natural gas consumption in 2011. (See Figure 3.12 and Figure 3.13.) The residential

FIGURE 3.11

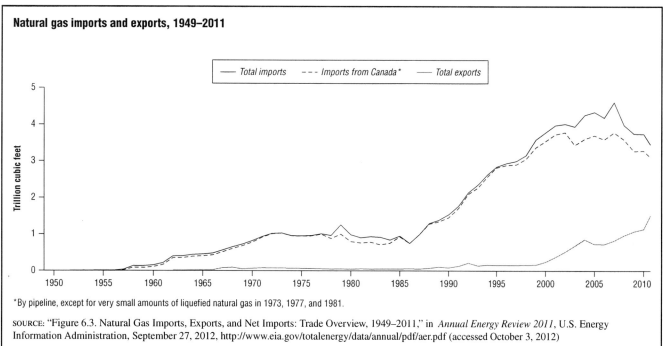

Natural gas imports and exports, 1949–2011

*By pipeline, except for very small amounts of liquefied natural gas in 1973, 1977, and 1981.

SOURCE: "Figure 6.3. Natural Gas Imports, Exports, and Net Imports: Trade Overview, 1949–2011," in *Annual Energy Review 2011*, U.S. Energy Information Administration, September 27, 2012, http://www.eia.gov/totalenergy/data/annual/pdf/aer.pdf (accessed October 3, 2012)

FIGURE 3.12

Natural gas consumption, by sector, 1949–2011

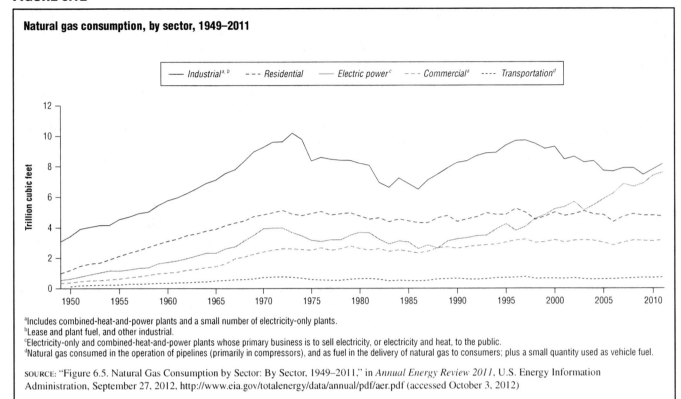

aIncludes combined-heat-and-power plants and a small number of electricity-only plants.
bLease and plant fuel, and other industrial.
cElectricity-only and combined-heat-and-power plants whose primary business is to sell electricity, or electricity and heat, to the public.
dNatural gas consumed in the operation of pipelines (primarily in compressors), and as fuel in the delivery of natural gas to consumers; plus a small quantity used as vehicle fuel.

SOURCE: "Figure 6.5. Natural Gas Consumption by Sector: By Sector, 1949–2011," in *Annual Energy Review 2011*, U.S. Energy Information Administration, September 27, 2012, http://www.eia.gov/totalenergy/data/annual/pdf/aer.pdf (accessed October 3, 2012)

sector used 4.7 tcf (133.9 bcm) and the commercial sector used 3.2 tcf (89.2 bcm). Usage by both sectors has been flat since the 1970s. In "Uses of Natural Gas," the EIA notes that slightly more than half of U.S. homes used natural gas as their primary heating fuel in 2012. Other residential uses include cooking, water heating, and clothes drying. Natural gas consumption for heating homes and commercial buildings depends heavily on weather conditions, in that colder winters result in increased demand. Residential consumption is also affected by conservation practices and the efficiency of gas appliances such as water heaters, stoves, and gas clothes dryers.

The transportation sector consumed only 720 bcf (20.4 bcm) of natural gas in 2011. (See Figure 3.12 and Figure 3.13.) According to the EIA, in *Annual Energy Review 2011*, consumption by this sector includes uses within the natural gas industry (e.g., during compression and delivery activities) and "a small quantity used as vehicle fuel."

WORLD NATURAL GAS PRODUCTION AND CONSUMPTION

Table 3.2 shows dry natural gas production for 1980, 1990, 2000, and 2010 for the world, by region, and for the top-25 producers. Total world production more than doubled between 1980 and 2010, from 53.4 tcf (1.5 tcm) to 111.8 tcf (3.2 tcm). Between 2000 and 2010 world

production increased by 28%. The Middle East experienced the largest growth, expanding its production by 120% during this period. North America has historically been the highest producer, but Eurasia has become a close competitor. The United States was the top-producing country in 2010, barely surpassing Russia. Canada, Iran, and Qatar finished out the top-five producers. U.S. production of 21.3 tcf (604.1 bcm) in 2010 accounted for 19% of the total world production.

Total world dry natural gas consumption is shown in Table 3.3. It grew from 52.9 tcf (1.5 tcm) in 1980 to 112.6 tcf (3.2 tcm) in 2010. Between 2000 and 2010 consumption grew by 29%. The Asia and Oceania region (95%) and the Middle East region (94%) experienced the highest growth surges between 2000 and 2010. Even so, the United States has historically been, by far, the leading consumer of dry natural gas among countries. U.S. consumption of 23.8 tcf (679.9 bcm) in 2010 accounted for 21% of the total world consumption. Other top consumers in 2010 included Russia, Iran, China, and Japan.

NATURAL GAS PRICES

Analysts typically track three types of natural gas prices: wellhead prices, city-gate prices, and consumer prices.

In *Annual Energy Review 2011*, the EIA explains that the wellhead is the point at which natural gas exits the ground. The wellhead price includes all production costs,

FIGURE 3.13

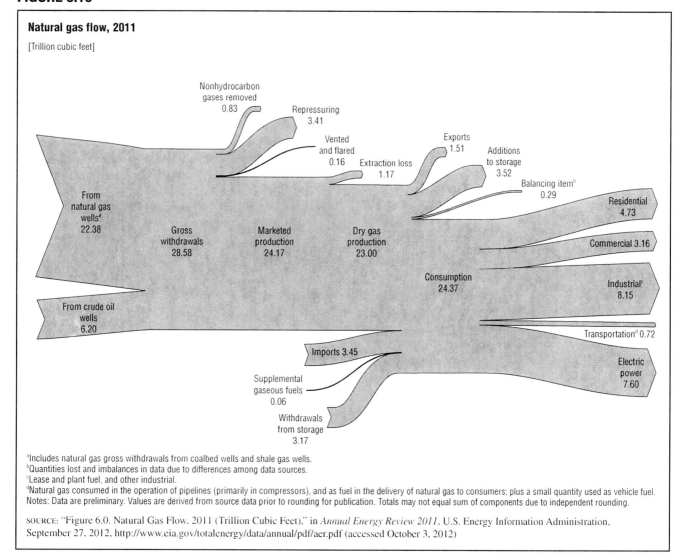

Natural gas flow, 2011

[Trillion cubic feet]

Nonhydrocarbon
gases removed
0.83

Repressuring
3.41

Vented
and flared
0.16

Exports
1.51

Additions
to storage
3.52

Extraction loss
1.17

Balancing item[b]
0.29

From
natural gas
wells[a]
22.38

Gross
withdrawals
28.58

Marketed
production
24.17

Dry gas
production
23.00

Residential
4.73

Commercial 3.16

Consumption
24.37

Industrial[c]
8.15

From crude oil
wells
6.20

Transportation[d] 0.72

Imports 3.45

Electric
power
7.60

Supplemental
gaseous fuels
0.06

Withdrawals
from storage
3.17

[a]Includes natural gas gross withdrawals from coalbed wells and shale gas wells.
[b]Quantities lost and imbalances in data due to differences among data sources.
[c]Lease and plant fuel, and other industrial.
[d]Natural gas consumed in the operation of pipelines (primarily in compressors), and as fuel in the delivery of natural gas to consumers; plus a small quantity used as vehicle fuel.
Notes: Data are preliminary. Values are derived from source data prior to rounding for publication. Totals may not equal sum of components due to independent rounding.

SOURCE: "Figure 6.0. Natural Gas Flow, 2011 (Trillion Cubic Feet)," in *Annual Energy Review 2011*, U.S. Energy Information Administration, September 27, 2012, http://www.eia.gov/totalenergy/data/annual/pdf/aer.pdf (accessed October 3, 2012)

such as gathering and compressing the gas, and state charges. The city-gate price is "a point or measuring station at which a distribution gas utility receives gas from a natural gas pipeline company or transmission system." Consumer prices are the prices paid by consumers to their local gas utilities. Natural gas prices vary across the nation because federal and state rate structures differ. Location also plays a role. For example, consumer prices are lower in major natural gas–producing areas where transmission costs are lower.

Table 3.4 lists wellhead, city-gate, and consumer prices, by sector, between 1973 and 2009 on an average annual basis and monthly and annually from January 2010 to June 2012. These are nominal prices, meaning that they reflect the prices at the time of purchase and have not been adjusted for inflation. Overall, residential customers paid the highest price ($11.02 per thousand cubic feet) for natural gas in 2011, followed by commercial customers ($8.93 per thousand cubic feet), industrial customers ($5.11 per thousand cubic feet), and electric power customers ($4.87 per thousand cubic feet).

Natural Gas Pricing Factors

Some of the factors described in Chapter 2 as affecting oil pricing, such as supply and demand, weather, and investor speculation, also impact natural gas pricing. However, because the United States produces nearly all the natural gas it consumes and obtains almost all of its imports from Canada, natural gas pricing in the United States is not dependent on foreign events and manipulation as is oil pricing.

The biggest impact on natural gas pricing within the United States in the 20th century was the deregulation of the natural gas industry, which began during the 1970s. As explained in Chapter 1, deregulation meant an end to government-protected monopolies and opened the industry to competition. These events and the subsequent restructuring of companies within the industry brought about a period of sharply rising prices. The EIA indicates in *Annual Energy Review 2011* that the real (inflation-adjusted) wellhead price climbed from around $0.70 per thousand cubic feet in 1970 to above $4 per thousand

TABLE 3.2

World production of dry natural gas, by region and selected country, selected years, 1980–2010

[Billion cubic feet]

	1980	1990	2000	2010	% change between 2000 and 2010
World	**53,375**	**73,788**	**87,100**	**111,845**	**28%**
North America	23,061	22,562	26,966	28,444	5%
Asia & Oceania	2,462	5,660	9,582	16,996	77%
Middle East	1,424	3,716	7,570	16,638	120%
Europe	9,144	8,596	10,982	10,830	−1%
Africa	686	2,459	4,440	7,377	66%
Eurasia	15,370	28,782	24,130	26,227	9%
Central & South America	1,227	2,014	3,430	5,333	55%
United States	19,403	17,810	19,182	21,332	11%
Russia	—	—	19,335	20,915	8%
Canada	2,759	3,849	6,470	5,390	−17%
Iran	250	837	2,127	5,161	143%
Qatar	184	276	1,028	4,121	301%
Norway	917	976	1,867	3,756	101%
China	505	508	962	3,334	246%
Netherlands	3,398	2,693	2,559	3,131	22%
Saudi Arabia	334	1,077	1,759	3,096	76%
Algeria	411	1,787	2,940	2,988	2%
Indonesia	654	1,602	2,237	2,917	30%
Malaysia	56	654	1,600	2,171	36%
Egypt	30	286	646	2,166	235%
Uzbekistan	—	—	1,992	2,123	7%
United Kingdom	1,323	1,754	3,826	1,988	−48%
India	51	399	795	1,848	133%
United Arab Emirates	200	780	1,355	1,811	34%
Australia	313	723	1,159	1,730	49%
Mexico	900	903	1,314	1,722	31%
Turkmenistan	—	—	1,642	1,600	−3%
Trinidad and Tobago	81	177	493	1,499	204%
Argentina	280	630	1,321	1,416	7%
Pakistan	286	482	856	1,400	64%
Thailand	—	208	658	1,281	95%
Nigeria	38	131	440	1,024	133%

SOURCE: Adapted from "Table. Dry Natural Gas Production (Billion Cubic Feet)," in *International Energy Statistics*, U.S. Energy Information Administration, 2012, http://www.eia.gov/cfapps/ipdbproject/iedindex3.cfm?tid=3&pid=26&aid=1&cid=regions&syid=1980&eyid=2011&unit=BCF (accessed October 7, 2012)

cubic feet during the early 1980s. However, once the industry adjusted to deregulation, prices dropped.

From the late 1980s through the late 1990s the well-head price was mostly in the range of $2 to $3 per thousand cubic feet. The price began to climb at the turn of the century and by 2008 it had reached a record high of $7.34 per thousand cubic feet. This time period coincides with the skyrocketing demand for natural gas from the electric power sector. (See Figure 3.12.) Because supply could not keep up with demand, natural gas prices rose. The industry responded with greater production, and prices began to come down. (See Figure 3.10.) By June 2012 the wellhead price was $2.54 per thousand cubic feet. (See Table 3.4.)

Chapter 2 explains various government interventions that are designed to benefit the oil and natural gas industry. According to the EIA, in *Direct Federal Financial Interventions and Subsidies in Energy in Fiscal Year 2010* (July 2011, http://www.eia.gov/analysis/requests/subsidy/pdf/subsidy.pdf), the natural gas and petroleum liquids industry received $2.8 billion in direct expenditures, tax breaks, research and development funds, and other means of financial support from the federal government in 2010. This was approximately 8% of the $37.2 billion that was allocated to the energy industry as a whole that year.

FUTURE TRENDS IN THE NATURAL GAS INDUSTRY

The EIA predicts that domestic natural gas production will greatly increase in the coming decades to keep up with rising demand. The agency believes that much of the new production will be gas obtained from unconventional geological sources—that is, shale gas and tight gas. (See Figure 3.14.) In fact, by 2035 shale gas is projected to supply nearly half of the nation's natural gas and tight gas to supply around 22%. All other sources will either decline or remain relatively constant in their production shares. The nation's major shale gas plays are the Marcellus play in the upper Appalachian basin, the Bakken shale play in North Dakota and Montana, and the Barnett play in Texas. (For a map of the shale plays in the U.S. mainland, see the EIA in "Lower 48 States Shale Plays" [2012, http://www.eia.gov/oil_gas/rpd/shale_gas.pdf].)

TABLE 3.3

World consumption of dry natural gas, by region and selected country, selected years, 1980–2010

[Billion cubic feet]

	1980	1990	2000	2010	% change between 2000 and 2010
World	52,943	73,629	87,259	112,607	29%
North America	22,559	22,470	27,723	28,845	4%
Eurasia	13,328	24,961	19,461	21,005	8%
Europe	11,193	13,360	17,394	20,638	19%
Asia & Oceania	2,577	5,865	10,517	20,466	95%
Middle East	1,311	3,599	6,822	13,245	94%
Central & South America	1,241	2,024	3,304	4,854	47%
Africa	735	1,351	2,038	3,554	74%
United States	19,877	19,174	23,333	23,775	2%
Russia	—	—	13,059	14,961	15%
Iran	232	837	2,221	5,106	130%
China	505	494	902	3,768	318%
Japan	903	2,028	2,914	3,718	28%
Germany	—	—	3,098	3,437	11%
United Kingdom	1,702	2,059	3,373	3,329	−1%
Saudi Arabia	334	1,077	1,759	3,096	76%
Canada	1,883	2,378	2,991	2,936	−2%
Italy	972	1,674	2,498	2,930	17%
India	51	399	795	2,277	187%
United Arab Emirates	105	663	1,110	2,138	93%
Mexico	799	918	1,398	2,135	53%
Netherlands	1,493	1,535	1,725	1,938	12%
Ukraine	—	—	2,779	1,877	−32%
France	981	997	1,403	1,699	21%
Egypt	30	286	646	1,630	152%
Uzbekistan	—	—	1,511	1,614	7%
Thailand	—	208	705	1,592	126%
Argentina	359	717	1,173	1,529	30%
Korea, South	—	107	669	1,515	127%
Indonesia	248	630	958	1,460	52%
Pakistan	286	482	856	1,400	64%
Turkey	—	122	524	1,346	157%
Spain	56	192	588	1,265	115%

SOURCE: Adapted from "Table. Dry Natural Gas Consumption (Billion Cubic Feet)," in *International Energy Statistics*, U.S. Energy Information Administration, 2012, http://www.eia.gov/cfapps/ipdbproject/iedindex3.cfm?tid=3&pid=26&aid=2&cid=regions&syid=1980&eyid=2011&unit=BCF (accessed October 7, 2012)

Substantial tight gas plays are found in the upper Appalachian basin, the Rocky Mountains, and the western portion of the Gulf Coast. (For a map of the tight gas plays in the U.S. mainland, see the EIA in "Major Tight Gas Plays, Lower 48 States" [2012, http://www.eia.gov/oil_gas/rpd/tight_gas.pdf].)

ENVIRONMENTAL ISSUES

The natural gas industry is associated with far fewer environmental concerns than are the oil and coal industries for two main reasons. First, leaks of natural gas (which do occur) do not contaminate ecosystems in the same way that oil spills do. Second, natural gas burns much cleaner than oil and coal. Nevertheless, leaks and the combustion of natural gas release large amounts of carbon into the atmosphere, which, as explained in Chapter 1, contributes to global warming and climate change.

The U.S. Environmental Protection Agency (EPA) publishes an annual report on the nation's emissions of greenhouse gases (gases that contribute to global warming). For comparison sake, the EPA converts the emission amounts of different gases from different sources into units of teragrams of carbon dioxide equivalent (Tg CO_2 Eq.). In *Inventory of U.S. Greenhouse Gas Emissions and Sinks: 1990–2010* (April 15, 2012, http://www.epa.gov/climatechange/Downloads/ghgemissions/US-GHG-Inventory-2012-Main-Text.pdf), the agency indicates that in 2010 fossil fuel combustion continued to account for the vast majority of greenhouse gas emissions, chiefly of carbon dioxide. Natural gas combustion contributed 1,261.6 Tg CO_2 Eq. (23%) of the total 5,387.8 Tg CO_2 Eq. of carbon dioxide emissions from fossil fuel combustion in 2010. This percentage was less than the percentage contributions of coal (36%) and petroleum (41%).

TABLE 3.4

Natural gas prices, by type and consuming sector, 1973–2009 annual and January 2010–June 2012 monthly

[Dollars per thousand cubic feet]

	Wellhead price	City-gate price	Consuming sectors								
			Residential		Commercial		Industrial		Transportation	Electric power	
			Price[a]	Percentage of sector	Price[a]	Percentage of sector	Price[a]	Percentage of sector	Vehicle fuel[b] Price[f]	Price[f]	Percentage of sector[c]
1973 average	0.22	NA	1.29	NA	0.94	NA	0.50	NA	NA	0.38	92.1
1975 average	0.44	NA	1.71	NA	1.35	NA	.96	NA	NA	0.77	96.1
1980 average	1.59	NA	3.68	NA	3.39	NA	2.56	NA	NA	2.27	96.9
1985 average	2.51	3.75	6.12	NA	5.50	NA	3.95	68.8	NA	3.55	94.0
1990 average	1.71	3.03	5.80	99.2	4.83	86.6	2.93	35.2	3.39	2.38	76.8
1995 average	1.55	2.78	6.06	99.0	5.05	76.7	2.71	24.5	3.98	2.02	71.4
1996 average	2.17	3.27	6.34	99.0	5.40	77.6	3.42	19.4	4.34	2.69	68.4
1997 average	2.32	3.66	6.94	98.8	5.80	70.8	3.59	18.1	4.44	2.78	68.0
1998 average	1.96	3.07	6.82	97.7	5.48	67.0	3.14	16.1	4.59	2.40	63.7
1999 average	2.19	3.10	6.69	95.2	5.33	66.1	3.12	18.8	4.34	2.62	58.3
2000 average	3.68	4.62	7.76	92.6	6.59	63.9	4.45	19.8	5.54	4.38	50.5
2001 average	4.00	5.72	9.63	92.4	8.43	66.0	5.24	20.8	6.60	4.61	40.2
2002 average	2.95	4.12	7.89	97.9	6.63	77.4	4.02	22.7	5.10	ᵃ3.68	83.9
2003 average	4.88	5.85	9.63	97.5	8.40	78.2	5.89	22.1	6.19	5.57	91.2
2004 average	5.46	6.65	10.75	97.7	9.43	78.0	6.53	23.6	7.16	6.11	89.8
2005 average	7.33	8.67	12.70	98.1	11.34	82.1	8.56	24.0	9.14	8.47	91.3
2006 average	6.39	8.61	13.73	98.1	12.00	80.8	7.87	23.4	8.72	7.11	93.4
2007 average	6.25	8.16	13.08	98.0	11.34	80.4	7.68	22.2	8.50	7.31	92.2
2008 average	7.97	9.18	13.89	97.5	12.23	79.9	9.65	20.5	11.75	9.26	101.1
2009 average	3.67	6.48	12.14	97.4	10.06	77.8	5.33	18.8	8.13	4.93	101.1
2010											
January	5.69	6.84	10.56	97.4	9.65	81.2	6.93	19.0	NA	6.98	101.0
February	5.30	6.64	10.69	97.8	9.71	81.8	6.76	18.6	NA	6.27	100.5
March	4.70	6.50	10.98	97.6	9.70	79.7	6.01	18.4	NA	5.47	101.0
April	4.10	5.88	11.97	96.2	9.55	75.7	5.12	17.7	NA	4.91	100.9
May	4.24	5.81	13.12	97.1	9.49	73.0	5.07	17.9	NA	4.96	100.9
June	4.27	6.02	14.86	96.9	9.73	71.9	5.03	18.0	NA	5.31	100.6
July	4.44	6.31	16.21	96.8	10.07	70.6	5.49	18.3	NA	5.34	100.6
August	4.38	6.22	16.65	96.4	9.96	69.8	5.37	17.8	NA	5.06	100.5
September	3.83	5.72	15.64	96.7	9.57	68.5	4.61	17.5	NA	4.61	100.7
October	4.05	5.70	13.37	96.8	9.28	71.8	4.74	16.8	NA	4.45	101.3
November	4.12	5.48	10.88	97.4	8.86	77.7	4.60	17.6	NA	4.55	101.0
December	4.68	5.74	9.88	97.4	8.56	80.2	5.42	17.8	NA	5.68	101.3
Average	**4.48**	**6.18**	**11.39**	**97.4**	**9.47**	**77.5**	**5.49**	**18.0**	**6.25**	**5.27**	**100.8**
2011											
January	ᴱ4.37	5.68	ᴿ9.89	ᴿ96.5	ᴿ8.76	ᴿ72.9	ᴿ5.63	ᴿ17.7	NA	5.63	101.5
February	ᴱ4.34	5.75	ᴿ10.13	ᴿ96.6	ᴿ8.90	ᴿ72.1	ᴿ5.75	ᴿ17.5	NA	5.28	102.1
March	ᴱ3.95	5.68	ᴿ10.33	ᴿ96.2	ᴿ8.91	ᴿ69.7	ᴿ5.19	ᴿ17.5	NA	4.82	101.2
April	ᴱ4.05	5.62	ᴿ11.26	ᴿ96.0	ᴿ9.04	ᴿ66.6	ᴿ5.33	ᴿ16.8	NA	5.03	101.8
May	ᴱ4.12	ᴿ5.79	ᴿ12.50	ᴿ96.2	ᴿ9.32	ᴿ64.1	ᴿ5.20	ᴿ17.2	NA	5.01	101.1
June	ᴱ4.20	ᴿ6.09	ᴿ14.67	ᴿ96.3	ᴿ9.53	ᴿ63.2	ᴿ5.19	ᴿ16.7	NA	5.19	101.2
July	ᴱ4.27	ᴿ6.15	ᴿ16.16	ᴿ96.3	ᴿ9.59	ᴿ59.4	ᴿ5.05	ᴿ17.5	NA	5.11	100.2
August	ᴱ4.20	6.19	ᴿ16.67	ᴿ95.7	ᴿ9.79	ᴿ58.2	ᴿ5.22	ᴿ16.9	NA	4.84	100.9
September	ᴱ3.82	5.93	ᴿ15.76	ᴿ95.6	ᴿ9.53	ᴿ57.9	ᴿ4.81	ᴿ16.6	NA	4.69	101.5
October	ᴱ3.62	5.43	ᴿ12.84	ᴿ95.6	ᴿ8.95	ᴿ58.3	ᴿ4.68	ᴿ16.7	NA	4.47	101.6
November	ᴱ3.35	ᴿ5.28	ᴿ10.79	ᴿ95.2	ᴿ8.64	ᴿ66.2	ᴿ4.63	ᴿ17.0	NA	4.24	101.2
December	ᴱ3.14	5.03	ᴿ9.84	ᴿ96.4	ᴿ8.33	ᴿ69.2	ᴿ4.57	ᴿ17.5	NA	4.15	101.4
Average	**ᴱ3.95**	**5.62**	**ᴿ11.02**	**ᴿ96.2**	**ᴿ8.93**	**ᴿ67.2**	**ᴿ5.11**	**ᴿ17.2**	**NA**	**4.87**	**101.2**
2012											
January	ᴱ2.89	4.86	ᴿ9.64	ᴿ96.2	ᴿ8.22	ᴿ70.4	ᴿ4.52	ᴿ16.9	NA	3.81	100.6
February	ᴱ2.46	ᴿ4.74	ᴿ9.51	ᴿ96.2	ᴿ7.94	ᴿ69.1	ᴿ4.30	ᴿ16.9	NA	3.45	100.5
March	ᴱ2.25	4.84	ᴿ10.45	ᴿ96.2	ᴿ8.40	ᴿ66.9	ᴿ3.69	ᴿ16.9	NA	3.07	100.2
April	ᴱ1.89	ᴿ4.20	ᴿ10.95	ᴿ95.6	ᴿ8.05	ᴿ63.2	ᴿ3.18	ᴿ16.2	NA	2.85	100.9
May	ᴱ1.94	ᴿ4.31	ᴿ12.46	ᴿ95.7	ᴿ7.93	ᴿ60.8	ᴿ2.99	ᴿ16.6	NA	3.02	100.9
June	ᴱ2.54	4.65	14.23	95.6	8.23	60.4	3.27	16.5	NA	3.20	100.7
6-month average	**ᴱ2.33**	**4.69**	**10.34**	**96.1**	**8.13**	**66.7**	**3.70**	**16.7**	**NA**	**3.21**	**100.7**
2011 6-month average	ᴱ4.17	5.73	10.62	96.4	8.96	69.7	5.39	17.3	NA	5.16	101.4
2010 6-month average	4.72	6.48	11.19	97.4	9.66	79.0	5.90	18.3	NA	5.64	100.8

TABLE 3.4

Natural gas prices, by type and consuming sector, 1973–2009 annual and January 2010–June 2012 monthly [CONTINUED]

[Dollars per thousand cubic feet]

[a]Includes taxes.
[b]Much of the natural gas delivered for vehicle fuel represents deliveries to fueling stations that are used primarily or exclusively by fleet vehicles. Thus, the prices are often those associated with the cost of gas in the operation of fleet vehicles.
[c]Percentages exceed 100 percent when reported natural gas receipts are greater than reported natural gas consumption—this can occur when combined-heat-and-power plants report fuel receipts related to non-electric generating activities.
R = Revised. NA = Not available. E = Estimate.
Notes: Prices are for natural gas, plus a small amount of supplemental gaseous fuels. Prices are intended to include all taxes. Wellhead annual and year-todate prices are simple averages of the monthly prices; all other annual and year-to-date prices are volume-weighted adverages of the monthly prices. Geographic coverage is the 50 states and the District of Columbia.

SOURCE: "Table 9.11. Natural Gas Prices (Dollars per Thousand Cubic Feet)," in *Monthly Energy Review: September 2012*, U.S. Energy Information Administration, September 26, 2012, http://www.eia.gov/totalenergy/data/monthly/archive/00351209.pdf (accessed October 3, 2012)

FIGURE 3.14

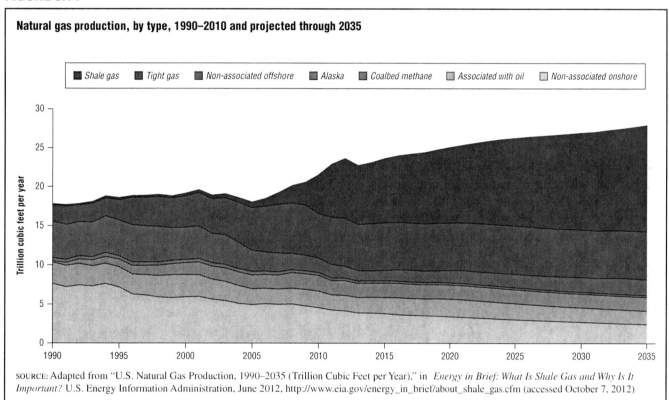

Natural gas production, by type, 1990–2010 and projected through 2035

SOURCE: Adapted from "U.S. Natural Gas Production, 1990–2035 (Trillion Cubic Feet per Year)," in *Energy in Brief: What Is Shale Gas and Why Is It Important?* U.S. Energy Information Administration, June 2012, http://www.eia.gov/energy_in_brief/about_shale_gas.cfm (accessed October 7, 2012)

CHAPTER 4
COAL

Even though it had been a source of energy for many centuries, coal was first used on a large scale during the Industrial Revolution (1760–1848). Its use expanded over the following decades with the building of railroads across the country. As noted in Chapter 1, coal replaced wood as the fuel for steam engines, which became the new workhorses of the modern age. However, coal's widespread use in industry waned during the latter 20th century as petroleum became more popular. Coal found a new niche in electricity generation, a field it now dominates. Coal's strong appeal is driven by its high abundance and easy accessibility in the United States. Nevertheless, coal has its problems, chiefly environmental ones.

UNDERSTANDING COAL

Coal is a dark, combustible, mineral solid. Coalbeds, or seams, are found in the earth between beds of sandstone, shale, and limestone. Like oil and thermogenic natural gas, coal developed underground over millions of years as organic (carbon-containing) material was subjected to intense temperatures and pressures. Oil and thermogenic natural gas originated from microscopic organisms, whereas coal is believed to have developed from vegetation, particularly plant fibers. These materials were first transformed into peat, a fibrous sod-like mass of partially disintegrated organic material. Peat is typically found near the ground surface and is very moist; however, once dried it can be burned as fuel. Peat created several millennia ago was baked and squeezed by the earth and eventually turned into coal.

Measuring Coal

In the United States coal amounts are measured in tons, which are also known as short tons. One ton is equivalent to 2,000 pounds. In countries that rely on the metric system, the metric tonne (t) is the preferred unit for measuring coal amounts.

Coal Ranks

There are different coal ranks (classifications), which are based on coal maturity. In *Annual Energy Review 2011* (September 2012, http://www.eia.gov/totalenergy/data/annual/pdf/aer.pdf), the Energy Information Administration (EIA) within the U.S. Department of Energy explains that coal ranks are evident by carbon and volatile matter contents, heat contents, and agglomeration (the tendency of particles to cake or stick together). Heat content is the amount of heat that can be produced from a given unit of fuel. In general, more mature coal contains less moisture than younger coal. Moisture content is very important because drier coals are easier to handle and burn hotter than wetter coals.

According to the EIA, the United States uses four coal ranks:

- Lignite (brown coal) is the lowest-ranked coal and the first true coal produced by the earth from peat. As such, it is relatively soft and moist compared with other coals. Lignite is brownish-black in color and has a high moisture content (as much as 45%). Lignite's heat content is about 9 million to 17 million British thermal units (Btu) per ton.

- Subbituminous coal ranks above lignite. Subbituminous coal ranges from dark brown to jet black in color and typically contains 20% to 30% moisture. Its heat content is approximately 17 million to 24 million Btu per ton.

- Bituminous coal is more mature than subbituminous coal and is the most abundant type of coal in the United States. Bituminous coal is dense and usually black, with a moisture content of less than 20%. It has a heat content of approximately 21 million to 30 million Btu per ton.

- Anthracite is the highest-ranked coal and the most mature. It is hard, brittle, and lustrous and typically jet-black in color, with a moisture content of less than 15%. Its heat content is approximately 22 million to 28 million Btu per ton. Anthracite is highly valued for its low moisture content and high heat content, but is very rare.

COAL DEPOSITS

The locations of coal deposits around the world are driven solely by geology. Regions and countries known to have large coal deposits include the United States, Eurasia (particularly China, Russia, and India), Australia, South Africa, and northern South America.

Geologists divide U.S. coalfields into the Appalachian, Interior, and Western regions. The Appalachian region encompasses the Appalachian Mountains in the eastern United States and stretches from Pennsylvania south to Alabama. The Interior region covers parts of the Midwest and deep South, including Texas. The Western region includes Alaska and a huge swath of territory from Montana and North Dakota in the north to the upper portions of Arizona and New Mexico in the south.

Like oil and natural gas, coal is located both onshore and offshore (i.e., beneath the earth's oceans). However, offshore coal deposits remain almost entirely untouched because of the enormous difficulties and costs involved in extracting them.

Estimates of the amounts of coal deposits around the world and in the United States that have not yet been extracted are provided in Chapter 7, along with information about coal exploration and development activities.

COAL EXTRACTION

Coal exists in the earth in solid form. As such, it must be mined out of the ground. The method used to mine coal depends on the terrain and the depth of the coal.

Surface mining, the most prevalent coal mining type, is used for deposits that lie relatively near the ground surface—that is, less than about 200 feet (61 m) deep. Surface mines can be developed in flat or hilly terrain. On large plots of relatively flat ground, workers use a technique known as area surface mining. Rock and soil that lie above the coal—called overburden or spoil—are loosened by drilling and blasting and then dug away. Another technique, contour surface mining, follows coal deposits along hillsides. Open pit mining—a combination of area and contour mining—is used to mine thick, steeply inclined coal deposits.

Underground mining is required when the coal lies more than 200 feet (61 m) below ground. The depth of most underground mines is less than 1,000 feet (305 m), but a few are 2,000 feet (610 m) deep. In underground mines, some coal must be left untouched to form pillars that prevent the mines from caving in.

There are three types of underground mines: a shaft mine, a slope mine, and a drift mine. In a shaft mine, elevators take miners and equipment up and down a vertical shaft to the coal deposit. By contrast, the entrance to a slope mine is an incline from the above-ground opening. In a drift mine, the mineshaft runs horizontally from the opening in the hillside. Workers and equipment are moved externally up the side of a hill or mountain to the mine entrance.

Before the early 1970s most U.S. coal was taken from underground mines. Since then, most coal production has shifted to surface mines.

Coal Mining Productivity

In the coal mining industry productivity is measured in terms of tons extracted per employee hour worked. Productivity has historically been higher for surface mining than for underground mining because surface mines are easier to work. (See Figure 4.1.) Surface mining is more common in the western United States than it is in the eastern United States. As a result, in 2011 productivity was higher at mines that were located west of the Mississippi River than at mines that were located east of the river. (See Figure 4.2.)

FIGURE 4.1

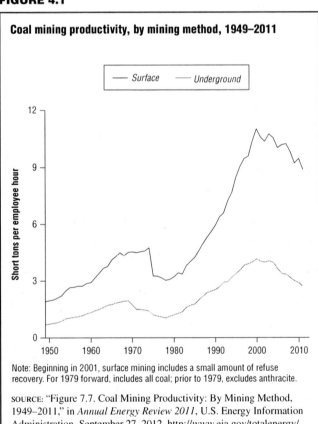

Coal mining productivity, by mining method, 1949–2011

— Surface   — Underground

Note: Beginning in 2001, surface mining includes a small amount of refuse recovery. For 1979 forward, includes all coal; prior to 1979, excludes anthracite.

SOURCE: "Figure 7.7. Coal Mining Productivity: By Mining Method, 1949–2011," in *Annual Energy Review 2011*, U.S. Energy Information Administration, September 27, 2012, http://www.eia.gov/totalenergy/data/annual/pdf/aer.pdf (accessed October 3, 2012)

FIGURE 4.2

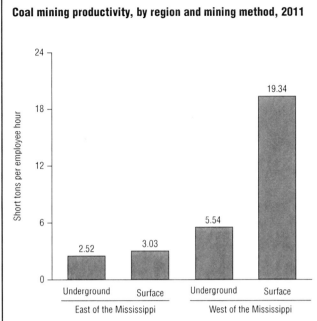

Coal mining productivity, by region and mining method, 2011

Note: Beginning in 2001, surface mining includes a small amount of refuse recovery.

SOURCE: "Figure 7.7. Coal Mining Productivity: By Region and Mining Method, 2011," in *Annual Energy Review 2011*, U.S. Energy Information Administration, September 27, 2012, http://www.eia.gov/totalenergy/data/annual/pdf/aer.pdf (accessed October 3, 2012)

PROCESSING AND TRANSPORTING COAL

Coal seams can contain noncoal materials, such as rock, ash, and other minerals. These impurities are removed at processing facilities (typically located at the mines) to meet the requirements of the coal buyers. One of the targeted impurities is sulfur, which is found mostly in the form of pyrite, an iron sulfide mineral. Impurity removal helps improve coal's heat content, lowers transportation costs, and makes the coal burn cleaner—that is, it generates less air pollutants, particularly sulfur dioxide. The coal may be crushed, either before or after cleaning, for easier handling.

Cleaned coal is transported from mine sites to customer locations. In the United States the primary transportation modes are freight trains and, to a lesser extent, barges and trucks. Crushed coal can also be mixed with water and piped as a slurry to its destination.

Coke

Some coal is subjected to a process called coking at specialized facilities. During coking, the coal is heated to a high temperature to eliminate volatile impurities and fuse together certain carbon and ash materials within the coal. According to the EIA, in *Annual Energy Review 2011*, bituminous coal with low ash and sulfur content is the only coal type used for coke manufacture. Coke has a high heat content of around 25 million Btu per ton and is used as a fuel and in certain industrial processes.

Waste Coal

Coal cleaning is not a perfect process; thus, some of the removed rock and mineral materials still contain enough hydrocarbons to make them useful combustible fuels. The EIA refers to these processing by-products as waste coal. In *Annual Energy Review 2011*, the agency lists several examples of waste coal with colorful names such as anthracite culm and bituminous gob. Technological advances in combustion equipment have made waste coal a marketable coal product.

Coal Synfuel

Coal can also be processed into fuels called coal synfuels. These may be either liquid or gaseous and can be manufactured by subjecting coal to high temperatures and pressures at industrial facilities.

Another method of synfuel production gaining in popularity is in situ (in place) coal gasification (ISCG) or underground coal gasification. ISCG processes burn coal while it is still in the ground to produce gas that can be pumped to the surface. Even though such technologies have been tried in the past, modern innovations have made ISCG a viable alternative for fuel production. It could conceivably produce huge amounts of energy from coal seams that are too deep to mine or otherwise untapped because of technological or economic reasons. Even though many ISCG demonstration projects were being conducted around the world as of December 2012, one of the most watched was the Swan Hills Synfuel project (http://swanhills-synfuels.com/) in Alberta, Canada, which is scheduled to begin commercial operation in 2015.

DOMESTIC COAL PRODUCTION

Traditional Coal

Figure 4.3 shows coal production between 1949 and 2011. During the 1950s and 1960s production hovered around 500 million tons (453.6 million t) per year. Demand for coal began building during the 1970s as a result of the Arab oil embargo, which was described in Chapter 2. Domestic coal production rose over subsequent decades, reaching 1 billion tons (907.2 million t) per year by the early 1990s. (See Figure 4.3.) Production has generally leveled off and even declined somewhat since then. In 2011 the United States produced almost 1.1 billion tons (992.7 million t) of coal. Table 4.1 provides a breakdown of production by rank, mining method, and general location.

Bituminous coal was the most produced coal type in the United States for several decades. (See Figure 4.4 and Table 4.1.) However, subbituminous coal began rising in production starting in the mid-1970s; by 2010 it surged past bituminous coal as the most produced coal type. In 2011 production by fuel type was:

- Subbituminous coal—510.5 million tons (463.1 million t)
- Bituminous coal—500.5 million tons (454 million t)

FIGURE 4.3

Coal production, consumption, and net exports, 1949–2011

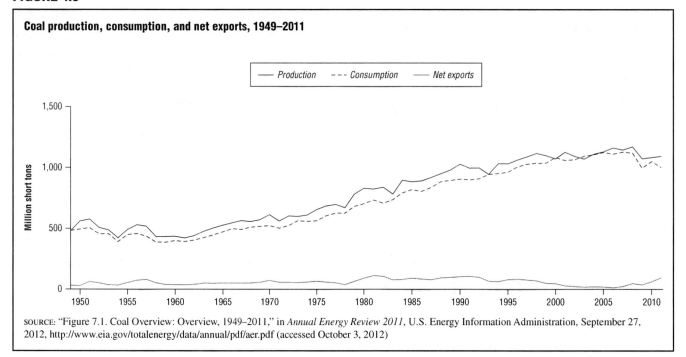

SOURCE: "Figure 7.1. Coal Overview: Overview, 1949–2011," in *Annual Energy Review 2011*, U.S. Energy Information Administration, September 27, 2012, http://www.eia.gov/totalenergy/data/annual/pdf/aer.pdf (accessed October 3, 2012)

FIGURE 4.4

Coal production, by coal type, 1949–2011

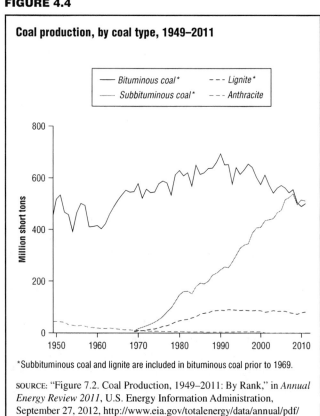

*Subbituminous coal and lignite are included in bituminous coal prior to 1969.

SOURCE: "Figure 7.2. Coal Production, 1949–2011: By Rank," in *Annual Energy Review 2011*, U.S. Energy Information Administration, September 27, 2012, http://www.eia.gov/totalenergy/data/annual/pdf/aer.pdf (accessed October 3, 2012)

• Lignite—81 million tons (73.5 million t)

• Anthracite—2.3 million tons (2.1 million t)

Figure 4.5 provides a production breakdown by mining method. Surface mining was the predominant method

for decades, but has become even more prevalent since the 1990s. In 2011, 748.8 million tons (679.3 million t) were surface mined, whereas 345.5 million tons (313.4 million t) were excavated via underground mining. Production by general location (i.e., either east or west of the Mississippi River) is shown in Figure 4.6.

Before 1999 most of the nation's coal was mined east of the Mississippi River. Miners had been digging deeper and deeper into the Appalachian Mountains for years before bulldozers began cutting open rich coal seams in the west. In 1965 western mines produced 27.4 million tons (24.9 million t), only 5% of the national total. (See Table 4.1.) By 1999 western production had increased more than 20-fold, to 570.8 million tons (517.8 million t), or 52% of the total. The amount of coal mined east of the Mississippi River that year was 529.6 million tons (480.4 million t). In 2011 mines west of the Mississippi River produced 638.5 million tons (579.2 million t)—58% of the total—whereas eastern mines produced 455.8 million tons (413.5 million t).

The growth in coal production in the western states resulted, in part, because of increased demand for low-sulfur coal, which is concentrated there. Low-sulfur coal burns cleaner than other coals and is less harmful to the environment. In addition, the coal is closer to the surface, so it can be extracted by surface mining, which is cheaper and more efficient than underground mining. Improved rail service has also made it easier to deliver low-sulfur coal mined in the western part of the country to electric power plants that are located in the eastern United States.

TABLE 4.1

Coal production, by type of coal, mining method, and location, selected years, 1949–2011

[Million short tons]

Year	Rank Bituminous coal[a]	Rank Subbituminous coal	Lignite	Anthracite[a]	Mining method Underground	Mining method Surface[a]	Location East of the Mississippi[a]	Location West of the Mississippi[a]	Total[a]
1949	437.9	[b]	[b]	42.7	358.9	121.7	444.2	36.4	480.6
1950	516.3	[b]	[b]	44.1	421.0	139.4	524.4	36.0	560.4
1955	464.6	[b]	[b]	26.2	358.0	132.9	464.2	26.6	490.8
1960	415.5	[b]	[b]	18.8	292.6	141.7	413.0	21.3	434.3
1965	512.1	[b]	[b]	14.9	338.0	189.0	499.5	27.4	527.0
1970	578.5	16.4	8.0	9.7	340.5	272.1	567.8	44.9	612.7
1975	577.5	51.1	19.8	6.2	293.5	361.2	543.7	110.9	654.6
1976	588.4	64.8	25.5	6.2	295.5	389.4	548.8	136.1	684.9
1977	581.0	82.1	28.2	5.9	266.6	430.6	533.3	163.9	697.2
1978	534.0	96.8	34.4	5.0	242.8	427.4	487.2	183.0	670.2
1979	612.3	121.5	42.5	4.8	320.9	460.2	559.7	221.4	781.1
1980	628.8	147.7	47.2	6.1	337.5	492.2	578.7	251.0	829.7
1981	608.0	159.7	50.7	5.4	316.5	507.3	553.9	269.9	823.8
1982	620.2	160.9	52.4	4.6	339.2	499.0	564.3	273.9	838.1
1983	568.6	151.0	58.3	4.1	300.4	481.7	507.4	274.7	782.1
1984	649.5	179.2	63.1	4.2	352.1	543.9	587.6	308.3	895.9
1985	613.9	192.7	72.4	4.7	350.8	532.8	558.7	324.9	883.6
1986	620.1	189.6	76.4	4.3	360.4	529.9	564.4	325.9	890.3
1987	636.6	200.2	78.4	3.6	372.9	545.9	581.9	336.8	918.8
1988	638.1	223.5	85.1	3.6	382.2	568.1	579.6	370.7	950.3
1989	659.8	231.2	86.4	3.3	393.8	586.9	599.0	381.7	980.7
1990	693.2	244.3	88.1	3.5	424.5	604.5	630.2	398.9	1,029.1
1991	650.7	255.3	86.5	3.4	407.2	588.8	591.3	404.7	996.0
1992	651.8	252.2	90.1	3.5	407.2	590.3	588.6	409.0	997.5
1993	576.7	274.9	89.5	4.3	351.1	594.4	516.2	429.2	945.4
1994	640.3	300.5	88.1	4.6	399.1	634.4	566.3	467.2	1,033.5
1995	613.8	328.0	86.5	4.7	396.2	636.7	544.2	488.7	1,033.0
1996	630.7	340.3	88.1	4.8	409.8	654.0	563.7	500.2	1,063.9
1997	653.8	345.1	86.3	4.7	420.7	669.3	579.4	510.6	1,089.9
1998	640.6	385.9	85.8	5.3	417.7	699.8	570.6	547.0	1,117.5
1999	601.7	406.7	87.2	4.8	391.8	708.6	529.6	570.8	1,100.4
2000	574.3	409.2	85.6	4.6	373.7	700.0	507.5	566.1	1,073.6
2001	[a]611.3	434.4	80.0	11.9	380.6	[a]747.1	[a]528.8	[a]598.9	[a]1,127.7
2002	572.1	438.4	82.5	1.4	357.4	736.9	492.9	601.4	1,094.3
2003	541.5	442.6	86.4	1.3	352.8	719.0	469.2	602.5	1,071.8
2004	561.5	465.4	83.5	1.7	367.6	744.5	484.8	627.3	1,112.1
2005	571.2	474.7	83.9	1.7	368.6	762.9	493.8	637.7	1,131.5
2006	561.6	515.3	84.2	1.5	359.0	803.7	490.8	672.0	1,162.7
2007	542.8	523.7	78.6	1.6	351.8	794.8	478.2	668.5	1,146.6
2008	555.3	539.1	75.7	1.7	357.1	814.7	493.3	678.5	1,171.8
2009	504.1	496.4	72.5	1.9	332.1	742.9	449.6	625.3	1,074.9
2010	[R]489.5	[R]514.8	[R]78.2	[R]1.8	[R]337.2	[R]747.2	[R]446.2	[R]638.2	[R]1,084.4
2011	[E]500.5	[E]510.5	[E]81.0	[E]2.3	[E]345.5	[E]748.8	[E]455.8	[E]638.5	[P]1,094.3

[a]Beginning in 2001, includes a small amount of refuse recovery.
[b]Included in "Bituminous Coal."
R = Revised. P = Preliminary. E = Estimate.
Note: Totals may not equal sum of components due to independent rounding.

SOURCE: "Table 7.2. Coal Production, Selected Years, 1949–2011 (Million Short Tons)," in *Annual Energy Review 2011,* U.S. Energy Information Administration, September 27, 2012, http://www.eia.gov/totalenergy/data/annual/pdf/aer.pdf (accessed October 3, 2012)

Coke

In *Annual Energy Review 2011*, the EIA indicates that coke production was 15.4 million tons (14 million t) in 2011. Coke production was down significantly compared with past decades. It peaked in 1955 at 75.3 million tons (68.3 million t), but steadily declined afterward, bottoming out at 11.1 million tons (10.1 million t) in 2009.

Waste Coal

The EIA estimates in *Annual Energy Review 2011* that 12.5 million tons (11.3 million t) of waste coal were supplied by the coal industry in 2011. This value was up from 1.4 million tons (1.3 million t) in 1989, when the agency first began tracking this product stream.

DOMESTIC COAL CONSUMPTION

Domestic coal consumption has been very close to domestic production since 1949. (See Figure 4.3.) Figure 4.7 shows domestic coal production, import, export, and consumption flows for 2011. Small amounts of coal were imported (13.1 million tons [11.9 million t]) and exported (107.3 million tons [97.3 million t]). In 2011 domestic consumption totaled just over 1 billion tons (907.2 million t).

FIGURE 4.5

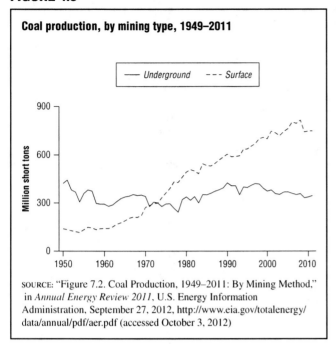

Coal production, by mining type, 1949–2011

— Underground --- Surface

SOURCE: "Figure 7.2. Coal Production, 1949–2011: By Mining Method," in *Annual Energy Review 2011*, U.S. Energy Information Administration, September 27, 2012, http://www.eia.gov/totalenergy/data/annual/pdf/aer.pdf (accessed October 3, 2012)

FIGURE 4.6

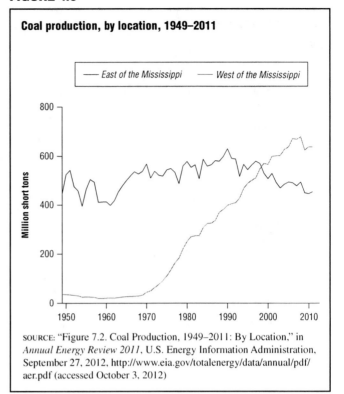

Coal production, by location, 1949–2011

— East of the Mississippi — West of the Mississippi

SOURCE: "Figure 7.2. Coal Production, 1949–2011: By Location," in *Annual Energy Review 2011*, U.S. Energy Information Administration, September 27, 2012, http://www.eia.gov/totalenergy/data/annual/pdf/aer.pdf (accessed October 3, 2012)

Coal Consumption by Sector

Figure 4.8 provides a historical breakdown of coal consumption by sector. Since the 1970s the electric power sector has been the dominant consumer of coal. In fact, in 2011 it used 928.6 million tons (842.4 million t) and accounted for 93% of the total consumption, up from only 17% in 1949. In the electric power sector, coal is pulverized and either burned directly or gasified. Steam from boilers or the hot coal gases turn turbines, which power generators that create electricity.

According to the EIA, in *Annual Energy Review 2011*, coal-fired plants produced 1.7 trillion kilowatt-hours of electricity in 2011, or 62% of U.S. electricity net generation. Net generation refers to the power available to the system—it does not include power used at the generating plants themselves.

The industrial sector was the second-largest coal consumer in 2011. (See Figure 4.8.) It used 71.7 million tons (65 million t), which was 7% of the total consumption, down from 44% in 1949. Coal is used in many industrial applications, including the chemical, cement, paper, synthetic fuels, metals, and food-processing industries. Coal was once a significant fuel source in the residential and commercial sectors. (See Figure 4.8.) After the 1940s, however, coal was replaced by oil, natural gas, and electricity. Thus, the residential and commercial sectors have become minor consumers of coal, as has the transportation sector.

Coke Consumption

The EIA notes in *Annual Energy Review 2011* that coke consumption was 15.8 million tons (14.3 million t)

in 2011. Nearly all (15.4 million tons [14 million t]) was supplied by domestic production. Net imports (imports minus exports) were 400,000 tons (363,000 t). Coke consumption has declined dramatically since the 1950s, when it was above 76 million tons (69 million t) per year.

Waste Coal Consumption

In *Annual Energy Review 2011*, the EIA indicates that waste coal consumption in 2011 was 12.5 million tons (11.3 million t), which was equal to waste coal supplied. The agency has been tracking waste coal amounts since 1989, when only 1.4 million tons (1.3 million t) of the materials were supplied and consumed domestically.

COAL IMPORTS AND EXPORTS

As noted earlier, domestic production and consumption have been closely matched for decades. However, the United States does import and export small amounts of coal. In *Annual Energy Review 2011*, the EIA indicates that net exports (exports minus imports) totaled 94.2 million tons (85.5 million t) in 2011. Nearly three-fourths (73%) of the imported coal came from Colombia. Smaller percentages were provided by Canada (13% of total), Indonesia (7% of total), and Venezuela (6% of total). The United States exported coal to numerous countries in 2011. The top destinations were the Netherlands (10% of total), Brazil (8% of total), the United Kingdom (6% of total), and Japan (6% of total).

FIGURE 4.7

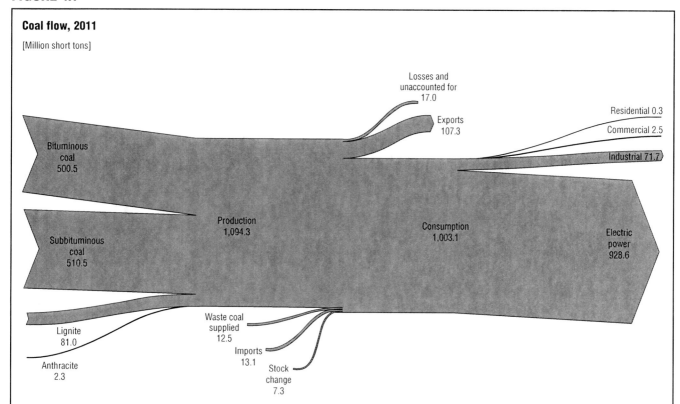

Coal flow, 2011

[Million short tons]

Losses and unaccounted for 17.0

Exports 107.3

Residential 0.3

Commercial 2.5

Industrial 71.7

Bituminous coal 500.5

Subbituminous coal 510.5

Production 1,094.3

Consumption 1,003.1

Electric power 928.6

Lignite 81.0

Anthracite 2.3

Waste coal supplied 12.5

Imports 13.1

Stock change 7.3

Notes: Production categories are estimated; all data are preliminary. Values are derived from source data prior to rounding for publication. Totals may not equal sum of components due to independent rounding.

SOURCE: "Figure 7.0. Coal Flow, 2011 (Million Short Tons)," in *Annual Energy Review 2011*, U.S. Energy Information Administration, September 27, 2012, http://www.eia.gov/totalenergy/data/annual/pdf/aer.pdf (accessed October 3, 2012)

FIGURE 4.8

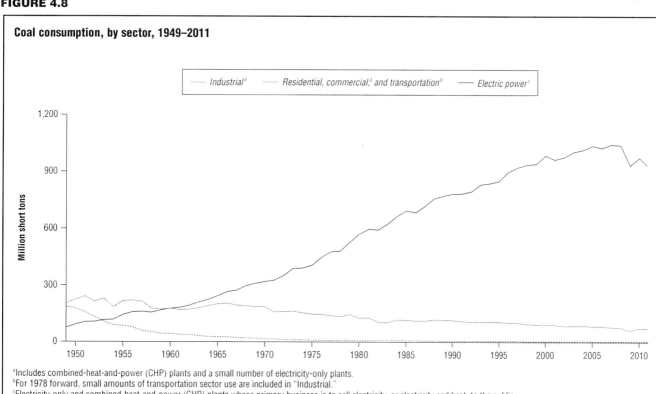

Coal consumption, by sector, 1949–2011

Industrial[a]   Residential, commercial,[a] and transportation[b]   Electric power[c]

[a]Includes combined-heat-and-power (CHP) plants and a small number of electricity-only plants.
[b]For 1978 forward, small amounts of transportation sector use are included in "Industrial."
[c]Electricity-only and combined-heat-and-power (CHP) plants whose primary business is to sell electricity, or electricity and heat, to the public.

SOURCE: "Figure 7.3. Coal Consumption by Sector: By Sector, 1949–2011," in *Annual Energy Review 2011*, U.S. Energy Information Administration, September 27, 2012, http://www.eia.gov/totalenergy/data/annual/pdf/aer.pdf (accessed October 3, 2012)

THE DOMESTIC OUTLOOK

In *Annual Energy Outlook 2012*, the EIA predicts that domestic coal production will decline through 2015 because of lower demand from the electric power sector as natural gas prices remain low. In addition, the coal industry is expected to face increasing competition from renewable energy sources for electricity generation. The EIA anticipates that stricter regulations on emissions, such as of sulfur dioxide and mercury, will force companies to retire older coal-fired power plants that cannot be economically outfitted with pollution-control technologies. These closures are expected to reduce coal demand through 2015; however, after that date the agency expects that growing demand for electricity, coupled with rising natural gas prices, will increase the demand for coal.

WORLD COAL PRODUCTION AND CONSUMPTION

Table 4.2 shows coal production for 1980, 1990, 2000, and 2010 for the world, by region, and for the top-25 producers. The total world coal production increased from 4.2 billion tons (3.8 billion t) in 1980 to 8 billion tons (7.2 billion t) in 2010. Between 2000 and 2010 world production increased by 63%. The Asia and Oceania region was the dominant regional producer in 2011, accounting for 5.1 billion tons (4.7 billion t), or 64% of the total world production. This region also experienced the highest production growth rate (139%) of any region between 2000 and 2010. China led all countries in coal production in 2010 with 3.5 billion tons (3.2 billion t), up 177% since 2000. The United States was second and India was third. Australia and Indonesia completed the top-five producers.

The total world coal consumption is shown in Table 4.3. It grew from 4.1 billion tons (3.7 billion t) in 1980 to 8 billion tons (7.3 billion t) in 2010. Between 2000 and 2010 world consumption grew by 59%. The Asia and Oceania region experienced the highest growth in consumption between 2000 and 2010, with an increase of 136%. China was, by far, the leading coal consumer in 2010; it consumed 3.7 billion tons (3.4 billion t), or 46% of the total world consumption. China's consumption

TABLE 4.2

World coal production, by region and selected country, selected years, 1980–2010

[Thousand short tons]

	1980	1990	2000	2010	% change between 2000 and 2010
World	**4,181,850**	**5,346,679**	**4,893,712**	**7,984,900**	**63%**
Asia & Oceania	1,037,513	1,791,201	2,153,576	5,136,770	139%
North America	874,145	1,112,986	1,162,355	1,171,117	1%
Europe	1,328,674	1,324,877	826,384	726,990	−12%
Eurasia	789,715	881,836	435,326	563,820	30%
Africa	138,739	201,708	255,855	285,839	12%
Central & South America	12,072	32,860	58,842	99,069	68%
Middle East	992	1,213	1,373	1,294	−6%
China	683,587	1,190,375	1,271,546	3,522,973	177%
United States	829,700	1,029,076	1,073,612	1,085,281	1%
India	125,845	247,569	370,018	622,818	68%
Australia	115,196	225,491	338,103	463,256	37%
Indonesia	627	11,628	84,541	370,379	338%
Russia	—	—	264,912	357,043	35%
South Africa	131,949	193,181	248,935	280,788	13%
Germany	—	—	226,048	200,955	−11%
Poland	253,517	237,082	179,247	146,237	−18%
Kazakhstan	—	—	85,367	122,135	43%
Colombia	4,533	22,562	42,044	81,957	95%
Turkey	20,776	52,280	69,741	79,090	13%
Canada	40,442	75,323	76,239	74,840	−2%
Greece	25,571	57,206	70,423	62,303	−12%
Czech Republic	—	—	71,829	60,941	−15%
Ukraine	—	—	68,788	60,117	−13%
Vietnam	5,732	5,113	12,797	49,233	285%
Serbia	—	—	—	41,169	N/A
Korea, North	48,619	51,029	32,786	34,785	6%
Romania	38,761	42,090	32,281	33,985	5%
Bulgaria	33,304	34,916	29,156	32,335	11%
Mongolia	5,274	7,889	5,715	27,831	387%
Thailand	1,620	13,732	19,605	20,346	4%
Estonia	—	—	12,927	19,769	53%
United Kingdom	143,758	104,055	33,731	19,640	−42%

SOURCE: Adapted from "Table. Total Primary Coal Production (Thousand Short Tons)," in *International Energy Statistics*, U.S. Energy Information Administration, 2012, http://www.eia.gov/cfapps/ipdbproject/iedindex3.cfm?tid=1&pid=7&aid=1&cid=regions&syid=1980&eyid=2011&unit=TST (accessed October 9, 2012)

TABLE 4.3

World coal consumption, by region and selected country, selected years, 1980–2010

[Thousand short tons]

	1980	1990	2000	2010	% change between 2000 and 2010
World	4,124,516	5,263,529	5,042,356	7,994,703	59%
Asia & Oceania	1,078,386	1,781,264	2,198,085	5,195,177	136%
North America	749,074	966,992	1,168,907	1,120,560	−4%
Europe	1,411,973	1,482,449	1,040,118	957,420	−8%
Eurasia	751,332	848,469	395,229	439,185	11%
Africa	113,252	152,283	189,056	217,208	15%
Central & South America	19,422	26,508	36,608	49,739	36%
Middle East	1,076	5,566	14,354	15,414	7%
China	678,511	1,124,129	1,239,407	3,695,378	198%
United States	702,730	904,498	1,084,095	1,048,295	−3%
India	122,929	247,864	403,408	721,986	79%
Russia	—	—	252,512	256,796	2%
Germany	—	—	269,815	255,746	−5%
South Africa	104,766	139,082	175,010	206,193	18%
Japan	97,571	126,610	168,798	205,983	22%
Poland	221,123	202,177	158,488	148,870	−6%
Australia	73,439	103,721	140,919	145,156	3%
Korea, South	30,766	49,774	71,057	125,575	77%
Turkey	22,600	59,990	88,845	109,120	23%
Kazakhstan	—	—	49,238	86,862	76%
Taiwan	6,589	18,936	51,957	71,375	37%
Ukraine	—	—	72,349	69,804	−4%
Greece	25,629	58,946	72,406	61,136	−16%
United Kingdom	133,560	119,377	65,627	55,374	−16%
Czech Republic	—	—	67,045	54,866	−18%
Indonesia	646	7,199	22,509	54,228	141%
Canada	41,321	53,849	69,616	52,498	−25%
Serbia	—	—	—	42,085	N/A
Thailand	1,620	14,228	24,174	38,948	61%
Romania	44,974	51,972	35,649	35,913	1%
Bulgaria	40,743	41,627	32,261	35,611	10%
Korea, North	49,335	53,675	32,709	30,350	−7%
Vietnam	5,190	4,608	8,607	25,719	199%

SOURCE: Adapted from "Table. Total Coal Consumption (Thousand Short Tons)," in *International Energy Statistics*, U.S. Energy Information Administration, 2012, http://www.eia.gov/cfapps/ipdbproject/iedindex3.cfm?tid=1&pid=1&aid=2&cid=regions&syid=1980&eyid=2011&unit=TST (accessed October 9, 2012)

grew 198% between 2000 and 2010. The United States was the second-largest consumer in 2010, followed by India, Russia, and Germany.

U.S. COAL PRICES

Table 4.4 lists coal prices by rank between 1949 and 2011. The prices are expressed in nominal dollars (current to that year) and in real dollars (inflation adjusted). The real dollar amounts assume that the value of a dollar was constant over time. Thus, changes in real prices reflect actual market variations and not inflationary effects. Real coal prices have experienced considerable up and down movement over the decades. (See Figure 4.9.) The price peaked in 1975 at nearly $58 per ton and then began to decline. Around the turn of the 21st century the price dipped to less than $19 per ton. It then grew slowly to around $33 per ton in 2011.

Figure 4.10 and Table 4.4 show price histories by coal rank. Anthracite has historically been the highest-priced type of coal. It sold for $70.99 per ton in 2011. Bituminous coal was priced at $57.64 per ton, followed

by lignite at $19.38 per ton and subbituminous coal at $15.80 per ton.

The EIA compares in *Annual Energy Review 2011* the production costs of coal, natural gas, and crude oil on a dollars per million Btu basis. The agency notes that at the point of first sale, coal cost $1.83 per million Btu in 2011. This compares with $3.60 per million Btu for natural gas (wellhead price) and $16.51 per million Btu for crude oil (domestic first purchase price). Thus, on a Btu basis, coal has been the least expensive fossil fuel since the late 1970s.

Coal Pricing Factors

The pricing for coal, like for oil and natural gas, is dependent on various supply and demand factors, weather, and investor speculation. Coal demand is particularly sensitive to weather conditions because the vast majority of production goes to electric power generation. Electricity demand varies strongly with outdoor temperature, given that heating and cooling homes and buildings are top electricity uses. Because the United States produces nearly all the coal it consumes, coal pricing in the

TABLE 4.4

Coal prices, by type, selected years, 1949–2011

[Dollars per short ton]

Year	Bituminous coal Nominal[b]	Bituminous coal Real[c]	Subbituminous coal Nominal[b]	Subbituminous coal Real[c]	Lignite[a] Nominal[b]	Lignite[a] Real[c]	Anthracite Nominal[b]	Anthracite Real[c]	Total Nominal[b]	Total Real[c]
1949	[d]4.90	[d][R]33.80	[d]	[d]	2.37	[R]16.35	8.90	[R]61.38	5.24	[R]36.14
1950	[d]4.86	[d][R]33.16	[d]	[d]	2.41	[R]16.44	9.34	[R]63.73	5.19	[R]35.41
1955	[d]4.51	[d][R]27.17	[d]	[d]	2.38	[R]14.34	8.00	[R]48.19	4.69	[R]28.25
1960	[d]4.71	[d][R]25.31	[d]	[d]	2.29	[R]12.30	8.01	[R]43.04	4.83	[R]25.95
1965	[d]4.45	[d][R]22.32	[d]	[d]	2.13	[R]10.68	8.51	[R]42.69	4.55	[R]22.82
1970	[d]6.30	[d][R]25.89	[d]	[d]	1.86	[R]7.64	11.03	[R]45.32	6.34	[R]26.05
1975	[d]19.79	[d][R]58.91	[d]	[d]	3.17	9.44	32.26	[R]96.04	19.35	[R]57.60
1976	[d]20.11	[d][R]56.62	[d]	[d]	3.74	[R]10.53	33.92	[R]95.50	19.56	[R]55.07
1977	[d]20.59	[d][R]54.50	[d]	[d]	4.03	[R]10.67	34.86	[R]92.26	19.95	[R]52.80
1978	[d]22.64	[d][R]55.99	[d]	[d]	5.68	[R]14.05	35.25	[R]87.18	21.86	[R]54.06
1979	27.31	[R]62.35	9.55	[R]21.80	6.48	[R]14.80	41.06	[R]93.75	23.75	[R]54.23
1980	29.17	[R]61.04	11.08	[R]23.18	7.60	[R]15.90	42.51	[R]88.95	24.65	[R]51.58
1981	31.51	[R]60.28	12.18	[R]23.30	8.85	[R]16.93	44.28	[R]84.71	26.40	[R]50.51
1982	32.15	[R]57.97	13.37	[R]24.11	9.79	[R]17.65	49.85	[R]89.89	27.25	[R]49.14
1983	31.11	[R]53.96	13.03	[R]22.60	9.91	[R]17.19	52.29	[R]90.70	25.98	[R]45.06
1984	30.63	[R]51.21	12.41	[R]20.75	10.45	[R]17.47	48.22	[R]80.61	25.61	[R]42.81
1985	30.78	[R]49.94	12.57	[R]20.40	10.68	[R]17.33	45.80	[R]74.32	25.20	[R]40.89
1986	28.84	[R]45.78	12.26	[R]19.46	10.64	[R]16.89	44.12	[R]70.04	23.79	[R]37.77
1987	28.19	[R]43.49	11.32	[R]17.46	10.85	[R]16.74	43.65	[R]67.34	23.07	[R]35.59
1988	27.66	[R]41.26	10.45	[R]15.59	10.06	[R]15.00	44.16	[R]65.87	22.07	[R]32.92
1989	27.40	[R]39.38	10.16	[R]14.60	9.91	[R]14.24	42.93	[R]61.70	21.82	[R]31.36
1990	27.43	[R]37.96	9.70	[R]13.42	10.13	[R]14.02	39.40	[R]54.52	21.76	[R]30.11
1991	27.49	[R]36.74	9.68	[R]12.94	10.89	[R]14.55	36.34	[R]48.57	21.49	[R]28.72
1992	26.78	[R]34.96	9.68	[R]12.64	10.81	[R]14.11	34.24	[R]44.70	21.03	[R]27.46
1993	26.15	[R]33.40	9.33	[R]11.92	11.11	[R]14.19	32.94	[R]42.07	19.85	[R]25.35
1994	25.68	[R]32.12	8.37	[R]10.47	10.77	[R]13.47	36.07	[R]45.12	19.41	[R]24.28
1995	25.56	[R]31.32	8.10	9.93	10.83	[R]13.27	39.78	[R]48.75	18.83	[R]23.07
1996	25.17	[R]30.27	7.87	[R]9.46	10.92	[R]13.13	36.78	[R]44.23	18.50	[R]22.25
1997	24.64	[R]29.12	7.42	[R]8.77	10.91	[R]12.89	35.12	[R]41.50	18.14	[R]21.43
1998	24.87	[R]29.06	6.96	[R]8.13	11.08	[R]12.95	42.91	[R]50.14	17.67	[R]20.65
1999	23.92	[R]27.54	6.87	[R]7.91	11.04	[R]12.71	35.13	[R]40.45	16.63	[R]19.15
2000	24.15	[R]27.22	7.12	[R]8.02	11.41	[R]12.86	40.90	[R]46.10	16.78	[R]18.91
2001	25.36	[R]27.95	6.67	[R]7.35	11.52	[R]12.70	47.67	[R]52.54	17.38	[R]19.16
2002	26.57	[R]28.82	7.34	[R]7.96	11.07	[R]12.01	47.78	[R]51.82	17.98	[R]19.50
2003	26.73	[R]28.40	7.73	8.21	11.20	11.90	49.87	[R]52.98	17.85	[R]18.96
2004	30.56	[R]31.57	8.12	8.39	12.27	12.68	39.77	[R]41.09	19.93	[R]20.59
2005	36.80	36.80	8.68	8.68	13.49	13.49	41.00	41.00	23.59	23.59
2006	39.32	[R]38.09	9.95	9.64	14.00	13.56	43.61	[R]42.25	25.16	24.37
2007	40.80	[R]38.41	10.69	10.06	14.89	[R]14.02	52.24	[R]49.18	26.20	[R]24.66
2008	51.39	[R]47.33	12.31	[R]11.34	16.50	[R]15.20	60.76	[R]55.96	31.25	[R]28.78
2009	55.44	[R]50.52	13.35	[R]12.17	17.26	[R]15.73	57.10	[R]52.04	33.24	[R]30.29
2010	[R]60.88	[R]54.85	[R]14.11	[R]12.71	[R]18.76	[R]16.90	[R]59.51	[R]53.62	[R]35.61	[R]32.08
2011[E]	57.64	50.85	15.80	13.94	19.38	17.10	70.99	62.62	36.91	32.56

[a]Because of withholding to protect company confidentiality, lignite prices exclude Texas for 1955–1977 and Montana for 1974–1978. As a result, lignite prices for 1974–1977 are for North Dakota only.
[b]Nominal costs represent costs at time of transaction.
[c]In chained (2005) dollars, calculated by using gross domestic product implicit price deflators.
[d]Through 1978, subbituminous coal is included in "Bituminous coal."
R = Revised. E = Estimate.
Note: Prices are free-on-board (F.O.B.) rail/barge prices, which are the F.O.B. prices of coal at the point of first sale, excluding freight or shipping and insurance costs. For 1949–2000, prices are for open market and captive coal sales; for 2001–2007, prices are for open market coal sales; for 2008 forward, prices are for open market and captive coal sales.

SOURCE: "Table 7.9. Coal Prices, Selected Years, 1949–2011 (Dollars per Short Ton)," in *Annual Energy Review 2011*, U.S. Energy Information Administration, September 27, 2012, http://www.eia.gov/totalenergy/data/annual/pdf/aer.pdf (accessed October 3, 2012)

United States is not dependent on foreign events and manipulation as is oil pricing.

As noted earlier, the electric power sector accounted for 93% of coal consumption in 2011. Thus, demand changes in that sector have a big impact on coal prices. Chapter 8 explains that natural gas is the chief competitor against coal in electricity production. When natural gas becomes cheaper than coal, the electric power sector uses more of its existing natural gas–burning units than its existing coal-burning units. This fuel switching greatly

reduces the demand for coal and puts downward pressure on coal prices.

Coal production and consumption have been relatively flat since the start of the 21st century. (See Figure 4.3.) However, coal prices have crept upward. Part of the upward pressure was due to increasing transportation costs. In "Cost of Transporting Coal to Power Plants Rose Almost 50% in Decade" (November 19, 2012, http://www.eia.gov/todayinenergy/detail.cfm?id=8830), the EIA indicates that the average cost of shipping coal

FIGURE 4.9

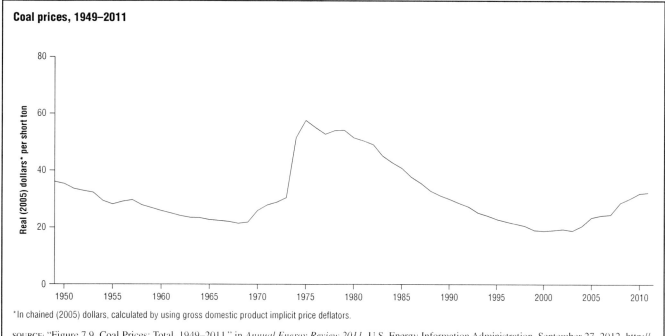

Coal prices, 1949–2011

*In chained (2005) dollars, calculated by using gross domestic product implicit price deflators.

SOURCE: "Figure 7.9. Coal Prices: Total, 1949–2011," in *Annual Energy Review 2011*, U.S. Energy Information Administration, September 27, 2012, http://www.eia.gov/totalenergy/data/annual/pdf/aer.pdf (accessed October 3, 2012)

FIGURE 4.10

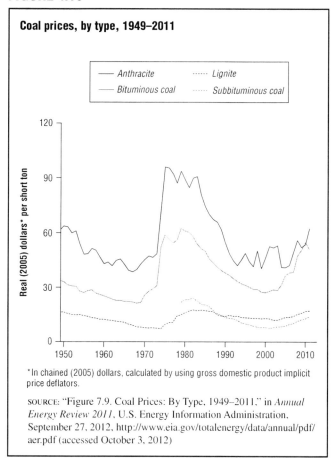

Coal prices, by type, 1949–2011

*In chained (2005) dollars, calculated by using gross domestic product implicit price deflators.

SOURCE: "Figure 7.9. Coal Prices: By Type, 1949–2011," in *Annual Energy Review 2011*, U.S. Energy Information Administration, September 27, 2012, http://www.eia.gov/totalenergy/data/annual/pdf/aer.pdf (accessed October 3, 2012)

by railroad to electric power plants increased by nearly 50% between 2001 and 2010. The United States' freight rail system is powered almost exclusively by diesel (a petroleum product), and as was discussed in Chapter 2, the price of oil skyrocketed during the first decade of the 21st century.

GOVERNMENT INTERVENTION. Coal pricing is also influenced by government actions. These include financial incentives, taxes, and regulations. Financial incentives put downward pressure on coal prices, whereas taxes and regulations push prices upward.

The EIA indicates in *Direct Federal Financial Interventions and Subsidies in Energy in Fiscal Year 2010* (July 2011, http://www.eia.gov/analysis/requests/subsidy/pdf/subsidy.pdf) that in 2010 the coal industry benefitted from nearly $1.4 billion in direct government expenditures, tax breaks, research and development funds, and other types of financial assistance from the federal government. Coal garnered less than 4% of the $37.2 billion in total energy-specific subsidies and support provided that year.

The federal government and some state governments impose various kinds of taxes on domestically produced coal. The collected monies often go into trust funds that help offset some of the external costs of coal mining. (As explained in Chapter 1, external costs are costs outside market costs.) The Black Lung Disability Fund has been operated since the late 1970s by the federal government and pays benefits under certain conditions to miners with black lung disease (a lung disease that is associated with coal mining). Federal and state taxes and fees also go into reclamation funds that are used to clean up abandoned mine sites.

ENVIRONMENTAL ISSUES

Environmental concerns related to the coal industry primarily involve the mining and combustion stages. In *Annual Coal Report 2011* (November 2012, http://www.eia.gov/coal/annual/pdf/acr.pdf), the EIA indicates that there were 1,325 coal mines in operation in 2011. Figure 4.11 shows the major mining areas of the country.

Coal is laden with heavy hydrocarbons and naturally occurring elements, such as mercury and sulfur. Combustion, even with pollution-control equipment, releases chemicals into the atmosphere that can negatively impact air and water quality, ecosystems, and human health. The EIA notes in *Annual Energy Review 2011* that coal burned for electricity generation and thermal output in 2010 resulted in emissions of 1.9 billion tons (1.7 billion t) of carbon dioxide, 4.9 million tons (4.5 million t) of sulfur dioxide, and 1.8 million tons (1.7 million t) of nitrogen oxides. The latter two gases are associated with air pollution problems, such as smog and acid rain (precipitation containing abnormally high levels of acids, particularly sulfuric acids). As explained in Chapter 1, carbon emissions are blamed for enhancing global warming and related climate changes. Emissions from coal-fired power plants can also contain mercury, a toxin that settles in water bodies and is absorbed by fish and other aquatic creatures that humans ingest.

Coal combustion is heavily regulated in the United States. Since 1970, when the Clean Air Act was passed, the government has imposed ever tighter restrictions on emissions and invested money in making coal a cleaner burning fuel. For example, in 1984 Congress established the Clean Coal Technology program and directed the Department of Energy to administer projects that were designed to demonstrate more environmentally friendly and economically efficient coal uses. Mechanical and chemical control measures have been developed through industry and government investment that have substantially

FIGURE 4.11

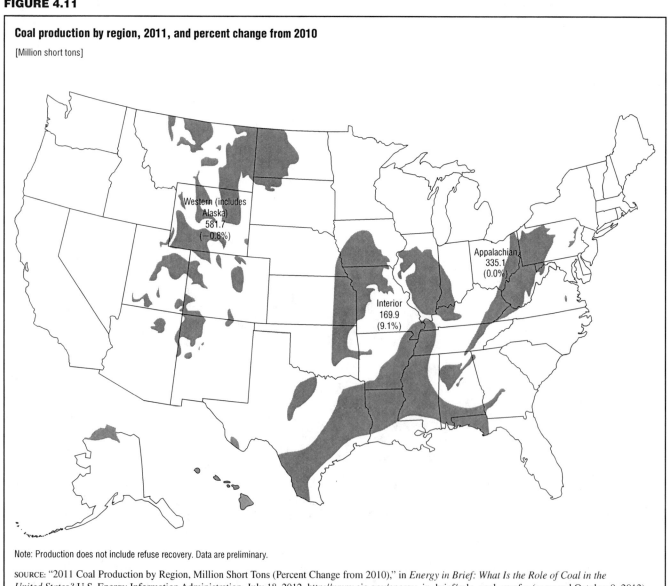

Coal production by region, 2011, and percent change from 2010

[Million short tons]

Western (includes Alaska)
581.7
(−0.8%)

Appalachian
335.1
(0.0%)

Interior
169.9
(9.1%)

Note: Production does not include refuse recovery. Data are preliminary.

SOURCE: "2011 Coal Production by Region, Million Short Tons (Percent Change from 2010)," in *Energy in Brief: What Is the Role of Coal in the United States?* U.S. Energy Information Administration, July 18, 2012, http://www.eia.gov/energy_in_brief/role_coal_us.cfm (accessed October 9, 2012)

reduced the emissions from coal combustion and improved the nation's air quality.

However, ever stricter standards have been implemented in the 21st century. For example, in 2011 the U.S. Environmental Protection Agency (EPA) finalized a rule requiring lower emissions from power plants. According to the EIA, in "Coal Plants without Scrubbers Account for a Majority of U.S. SO_2 Emissions" (December 21, 2011, http://www.eia.gov/todayinenergy/detail.cfm?id=4410), the Cross-State Air Pollution Rule calls for the electric power sector to reduce its sulfur dioxide emissions by 53% by 2014. In March 2012 the EPA (http://epa.gov/carbonpollu tionstandard/actions.html) proposed implementing a new type of regulation that would limit carbon emissions from power plants. The limits would apply only to new plants that begin construction at least one year after the regulation is finalized. As of December 2012, the regulation had not been finalized. More information about pollution-control regulations and global warming mitigation techniques (such as carbon sequestration), which are related to electric power generation from coal, is included in Chapter 8.

CHAPTER 5
NUCLEAR ENERGY

During the early 20th century scientists succeeded in releasing energy that was bound up in the atom. Using a process called fission, they split apart the nucleus of a heavy atom (an atom containing many protons and neutrons) into two lighter nuclei. (See Figure 5.1.) The two resulting nuclei contained less mass than the original nucleus because some of the original atomic mass was converted into energy in the form of heat and radiation. Scientists knew that the key to harnessing this energy was setting up a chain reaction in which numerous heavy nuclei could be split apart in a confined space under controlled conditions. In 1942 the physicist Enrico Fermi (1901–1954) achieved this feat at the University of Chicago in Illinois by creating the first self-sustaining nuclear fission chain reaction. His work transformed energy production as the techniques were quickly refined to produce electricity using the heat that was generated during controlled fission reactions.

At first, nuclear power was hailed as a super energy source that could provide huge amounts of electricity without the need for burning air-polluting fossil fuels. However, nuclear power's potential has been tempered by the logistical, environmental, and safety considerations involved. Controlling chain reactions and disposing of the radioactive materials that result from them have proven to be massive challenges.

UNDERSTANDING NUCLEAR ENERGY

Radiation is a form of energy transfer that naturally results from the spontaneous emission of energy and/or high-energy particles from the nucleus of an atom. The earth is bombarded by radiation from the sun and other sources in outer space. In addition, many naturally radioactive elements, such as uranium, are found within the earth's crust and oceans. Scientists first produced nuclear energy by bombarding the nuclei of an isotope of uranium called uranium 235 (U-235). (Isotopes are atoms of an element that have the usual number of protons but different numbers of neutrons in their nuclei.)

Under proper conditions, the fission of a U-235 atom creates the needed cascading chain of nuclear reactions. If this series of reactions is regulated to occur slowly, as it is in nuclear power plants, the energy emitted can be captured for generating electricity. If this series of reactions is allowed to occur all at once, as in an atomic bomb, the energy emitted is explosive. (Plutonium 239 can also be used to generate a chain reaction similar to that of U-235.)

Figure 5.2 shows a pressurized-water nuclear reactor, the most common type of reactor in commercial use. Bombardment of uranium in the core of the reactor (shown as "1" in Figure 5.2) generates a nuclear reaction (fission) that produces heat. The heat from the reaction is carried away by water under high pressure ("2") to a steam generator ("3"). This heat vaporizes the water in the steam generator, producing steam. The steam is carried by a steamline ("4") to a turbine, making the attached generator spin, which produces electricity. The large cooling towers that are associated with nuclear plants cool the steam after it has run through the turbines. A boiling-water reactor works much the same way, except that the water surrounding the core boils and directly produces the steam, which is then piped to the turbine generator.

Commercial nuclear power reactors exclusively produce electrical power, which is measured by a basic unit called a watt. Because the watt is a relatively small unit, power plant capacities are typically measured in a multiple of watts, such as the kilowatt (10^3 watts), megawatt (10^6 watts), gigawatt (10^9), or terawatt (10^{12} watts). Electrical power production is often expressed in multiples of a kilowatt-hour (the amount of work done by a kilowatt over a period of one hour).

FIGURE 5.1

Nuclear chain reaction

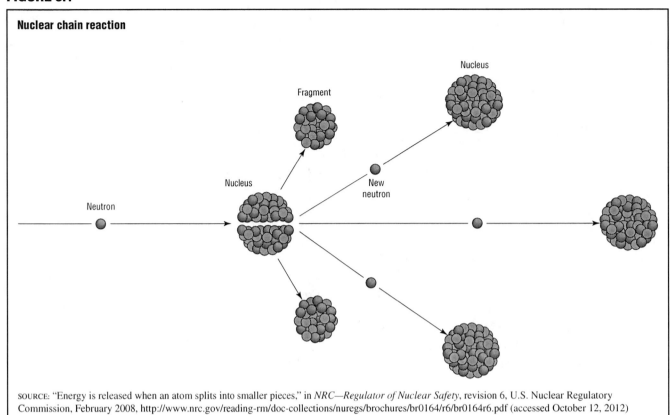

SOURCE: "Energy is released when an atom splits into smaller pieces," in *NRC—Regulator of Nuclear Safety*, revision 6, U.S. Nuclear Regulatory Commission, February 2008, http://www.nrc.gov/reading-rm/doc-collections/nuregs/brochures/br0164/r6/br0164r6.pdf (accessed October 12, 2012)

DOMESTIC URANIUM PRODUCTION AND PRICING

Uranium is present in low concentrations throughout the earth's soils, rocks, and water bodies. It is most highly concentrated in certain rock and mineral deposits called uranium ore. In "Where Our Uranium Comes From" (July 2, 2012, http://www.eia.gov/energyexplained/index.cfm?page=nuclear_where), the Energy Information Administration (EIA) within the U.S. Department of Energy (DOE) explains that domestic uranium ore deposits are located primarily in the western United States. The uranium in these deposits can be recovered by mining the ore or through in situ (in place) leaching. In the latter method, a chemical solution is pumped through the underground ore deposits, and the extracted uranium is pumped to the surface for processing.

For decades, mining uranium ore was the most common uranium recovery method used in the United States. The U.S. Environmental Protection Agency indicates in *Uranium Location Database Compilation* (August 2006, http://www.epa.gov/radiation/docs/tenorm/402-r-05-009.pdf) that from the 1940s through the 1990s thousands of uranium mines were operated, mostly in Arizona, Colorado, New Mexico, Utah, and Wyoming. Uranium ore is mined using methods that are similar to those used for other metal ores. A major difference is that uranium mining exposes workers to radioactivity. Uranium atoms

naturally split by themselves at a slow rate, causing radioactive substances such as radon to accumulate slowly in the deposits.

According to the EIA, in *2011 Domestic Uranium Production Report* (May 2012, http://www.eia.gov/uranium/production/annual/pdf/dupr.pdf), only 10 uranium mines were in operation in the United States in 2011. Five of the mines produced ore and the other five produced uranium derived from in situ leaching. Another facility produced uranium using an undisclosed method. Figure 5.3 shows a third uranium recovery option called heap leaching. In *U.S. Nuclear Regulatory Commission Information Digest 2012–2013* (August 2012, http://www.nrc.gov/reading-rm/doc-collections/nuregs/staff/sr1350/v24/sr1350v24.pdf), Ivonne Couret et al. describe heap leaching as a process in which acid is dripped over aboveground piles (heaps) of uranium ore to separate the uranium. Couret et al. note that as of 2012 the U.S. Nuclear Regulatory Commission (NRC) did not license any heap leach facilities; however, the commission expected to begin receiving applications for them in coming years.

The EIA explains in *Annual Energy Review 2011* (September 2012, http://www.eia.gov/totalenergy/data/annual/pdf/aer.pdf) that once mined, uranium ore is shipped to milling or processing facilities and processed into a powder called uranium concentrate. In situ leaching

FIGURE 5.2

Pressurized-water reactor

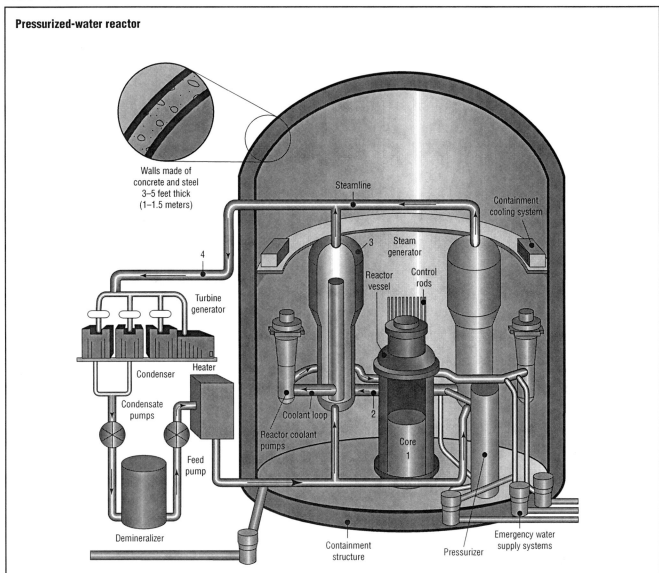

SOURCE: Ivonne Couret et al., "Figure 15. Typical Pressurized-Water Reactor," in *U.S. Nuclear Regulatory Commission Information Digest 2012–2013*, vol. 24, U.S. Nuclear Regulatory Commission, Office of Public Affairs, August 2012, http://www.nrc.gov/reading-rm/doc-collections/nuregs/staff/sr1350/v24/sr1350v24.pdf (accessed October 14, 2012)

facilities produce concentrate from the solutions they extract from the ground.

Uranium concentrate has the chemical formula U_3O_8 (uranium oxide) and is commonly called yellowcake, although it may be yellow or brown in color. Domestic uranium oxide production peaked in 1980 at 43.7 million pounds (19.8 million kg). (See Figure 5.4.) Since that time domestic production has plummeted. It was 4 million pounds (1.8 million kg) in 2011. The United States has increasingly relied on imports to meet its uranium demand. In 2011, 54 million pounds (24.5 million kg) of uranium oxide were imported.

According to the EIA, in *2011 Uranium Marketing Annual Report* (May 2012, http://www.eia.gov/uranium/marketing/pdf/2011umar.pdf). in 2011 the United States

imported uranium from 17 countries. The major suppliers included Australia, Canada, Kazakhstan, Namibia, and Russia. As shown in Figure 5.4, the United States also exports uranium oxide. In 2011, 16.7 million pounds (7.6 million kg) were exported.

Yellowcake is a feedstock to the nuclear fuel cycle and undergoes significant processing. First, it is converted into uranium hexafluoride (UF_6) by combining uranium with fluorine gas. Cylinders of this gas are sent to a gaseous diffusion plant, where the uranium is heated in a furnace (UF_6 vaporization). (See "1" in Figure 5.5 and "Conversion" in Figure 5.3.) Through this process the uranium is enriched (made more concentrated) and is then converted into oxide powder (UO_2). The oxide powder is made into fingertip-sized fuel pellets. (See Figure 5.6.) The pellets are stacked in rods, which are tubes about

FIGURE 5.3

Nuclear fuel cycle

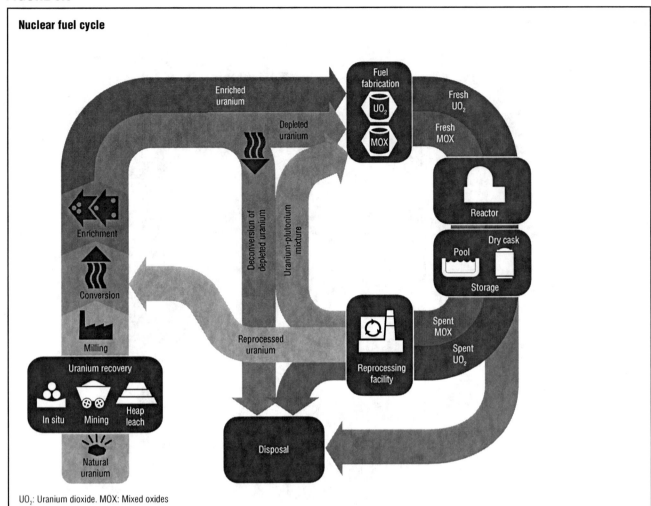

UO$_2$: Uranium dioxide. MOX: Mixed oxides

SOURCE: Ivonne Couret et al., "Nuclear Fuel Cycle," in *U.S. Nuclear Regulatory Commission Information Digest 2012–2013*, vol. 24, U.S. Nuclear Regulatory Commission, Office of Public Affairs, August 2012, http://www.nrc.gov/reading-rm/doc-collections/nuregs/staff/sr1350/v24/sr1350v24.pdf (accessed October 14, 2012)

12 feet (3.7 m) long. Many rods (see "control rods" in Figure 5.2 and "fuel rods" in Figure 5.6) are bundled together in assemblies, and hundreds of these assemblies make up the core of a nuclear reactor.

As shown in Figure 5.5, some of the uranium that is "left over" from powder processing/pellet manufacturing is recycled. Waste that cannot be recycled is shipped to a nuclear waste facility.

It should be noted that Figure 5.3 also shows mixed oxide (MOX) fuel being fed to a nuclear reactor. MOX contains a mixture of oxides, typically uranium mixed with plutonium. As of December 2012, the United States did not have an MOX facility in operation that produced the fuel mixture for commercial nuclear power reactors. However, a facility was being constructed at the DOE's Savannah River Site near Augusta, Georgia. In "The MOX Project" (February 27, 2012, http://nuclear.duke-energy.com/category/mox), the power company Duke Energy explains that the MOX produced at the facility will be approximately 95% uranium oxide and 5% plutonium oxide. The latter will be obtained from old nuclear warheads that the United States has been dismantling. The facility is scheduled to begin operations in 2016.

Uranium Prices

Figure 5.7 shows the average prices for domestically produced and imported uranium oxide in dollars per pound between 1981 and 2011. Note that the prices are not adjusted for inflation. During the early 1980s the price of domestically produced uranium was nearly $40 per pound. However, by the early to mid-1990s the price was closer to $10 per pound and remained near this level through the turn of the century before beginning to rise. By 2011 the price of uranium oxide, both domestically produced and imported, was around $54 per pound. The dramatic increase was caused, in large part, by increased demand for uranium from nuclear power producers. According to the EIA, in *Annual Energy Review 2011*, nuclear electric plants built up huge stockpiles of uranium oxide during the 1980s. These inventories had dropped dramatically by the start of the 21st century, prompting plants to increase their purchases.

DOMESTIC NUCLEAR ENERGY PRODUCTION

Figure 5.8 shows the number of operable nuclear generating units in the United States. The number peaked in 1990 at 112 units and then declined. As of August 2012, 104 nuclear generating units were operable. According to Couret et al., the units were located in 31 states, but heavily concentrated in the eastern United States. Most (69) of the reactors were the pressurized-water type (the type illustrated in Figure 5.2). The remaining 35 reactors were boiling-water reactors.

The EIA notes in *Annual Energy Review 2011* that between 1980 and 2011 no new construction permits were issued for nuclear power plants in the United States. In addition, some plants were permanently shut down. The decline in nuclear power plants stems from several related issues. Financing is difficult to find, and construction has become more expensive, partly because of longer delays for licensing, but also because of regulations that were instituted following an accident at the Three Mile Island nuclear power plant in Pennsylvania in 1979. As will be explained later in this chapter, that accident allowed nuclear fuel to overheat to a dangerous level. No large-scale releases or explosions resulted; however, much stricter controls were put into place to regulate nuclear power plant operations.

Even though no new nuclear power plants have been built for decades, the output of electricity at existing plants has increased. The net summer capacity of operable units increased from 51.8 million kilowatts in 1980 to 101.4 million kilowatts in 2011. (See Figure 5.9.) A capacity factor is the proportion of electricity produced compared to what could have been produced at full-power operation. The EIA reveals in *Annual Energy Review 2011* that in 2011 the average capacity factor for U.S. nuclear power plants was 89.1%, near the all-time high of 91.8% set in 2007, and far greater than the value of 56.3% reported in 1980. Better training for operators, longer operating cycles between refueling,

FIGURE 5.4

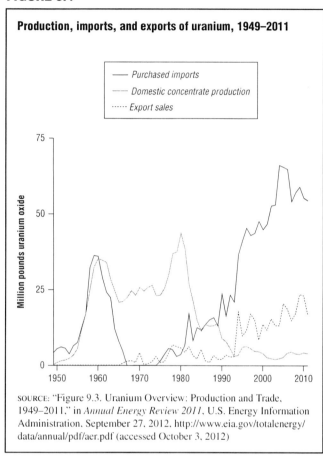

Production, imports, and exports of uranium, 1949–2011

— Purchased imports
--- Domestic concentrate production
······ Export sales

Million pounds uranium oxide

SOURCE: "Figure 9.3. Uranium Overview: Production and Trade, 1949–2011," in *Annual Energy Review 2011*, U.S. Energy Information Administration, September 27, 2012, http://www.eia.gov/totalenergy/data/annual/pdf/aer.pdf (accessed October 3, 2012)

FIGURE 5.5

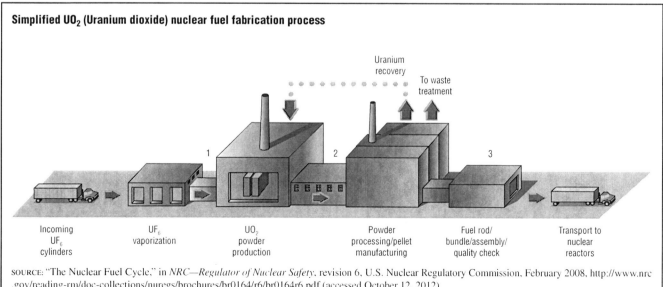

Simplified UO₂ (Uranium dioxide) nuclear fuel fabrication process

Uranium recovery

To waste treatment

| 1 | 2 | 3 |

Incoming UF₆ cylinders

UF₆ vaporization

UO₂ powder production

Powder processing/pellet manufacturing

Fuel rod/ bundle/assembly/ quality check

Transport to nuclear reactors

SOURCE: "The Nuclear Fuel Cycle," in *NRC—Regulator of Nuclear Safety*, revision 6, U.S. Nuclear Regulatory Commission, February 2008, http://www.nrc.gov/reading-rm/doc-collections/nuregs/brochures/br0164/r6/br0164r6.pdf (accessed October 12, 2012)

FIGURE 5.6

Fuel rod containing uranium fuel pellets

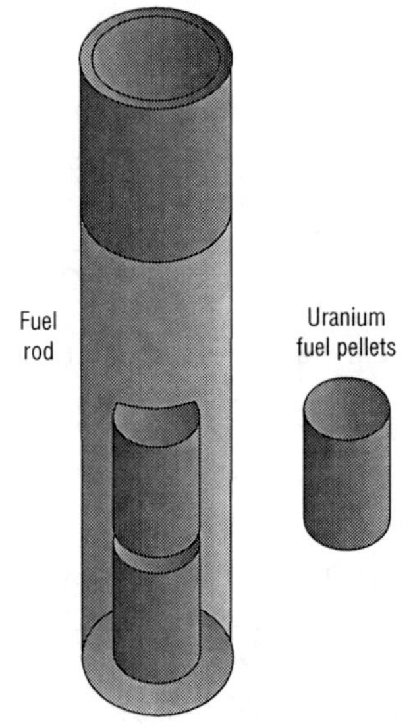

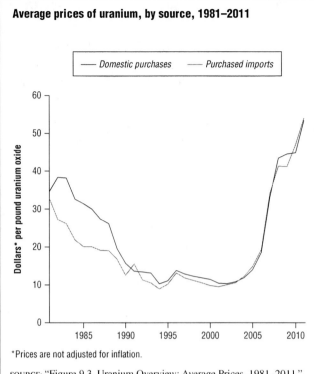

SOURCE: Adapted from Ivonne Couret et al., "Figure 41. Spent Fuel Generation and Storage after Use," in *U.S. Nuclear Regulatory Commission Information Digest 2012–2013*, vol. 24, U.S. Nuclear Regulatory Commission, Office of Public Affairs, August 2012, http://www.nrc.gov/reading-rm/doc-collections/nuregs/staff/sr1350/v24/sr1350v24.pdf (accessed October 14, 2012)

FIGURE 5.7

Average prices of uranium, by source, 1981–2011

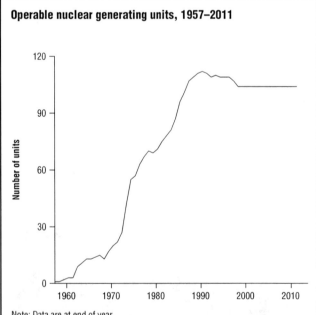

*Prices are not adjusted for inflation.

SOURCE: "Figure 9.3. Uranium Overview: Average Prices, 1981–2011," in *Annual Energy Review 2011*, U.S. Energy Information Administration, September 27, 2012, http://www.eia.gov/totalenergy/data/annual/pdf/aer.pdf (accessed October 3, 2012)

and control-system improvements have contributed to increased plant performance and an increase in the capacity factor.

The percentage of U.S. electricity supplied by nuclear power grew considerably during the 1970s and early to mid-1980s and then leveled off. According to the EIA, nuclear power supplied only 4.5% of the total electricity generated in the United States in 1973. In 2011 nuclear electricity net generation was 790.2 billion kilowatt-hours (kWh). (See Figure 5.10.) This was 19.2% of the nation's total electricity generation of 4,105.7 billion kWh. According to the EIA, nuclear power's share has hovered around 19% to 20% since 1990 and climbed as high as 20.6% in 2001.

WORLD NUCLEAR POWER PRODUCTION

As shown in Table 5.1, world nuclear power production totaled 2.5 million gigawatt-hours (GWh) in 2011. Thirty-one countries had 436 units in operation at that time. The United States, by far, had the most units (104) and the highest nuclear power production (790,439 GWh). Other countries with substantial production included France with 423,509 GWh from 58 units, Russia with 162,018 GWh from 33 units, and Japan with 156,182 GWh from 50 units.

Figure 5.11 lists the nuclear share of electricity that was generated by all the countries except Iran in 2011.

FIGURE 5.8

Operable nuclear generating units, 1957–2011

Note: Data are at end of year.
Units holding full-power operating licenses, or equivalent permission to operate, at the end of the year.

SOURCE: "Figure 9.1. Nuclear Generating Units: Operable Units, 1957–2011," in *Annual Energy Review 2011*, U.S. Energy Information Administration, September 27, 2012, http://www.eia.gov/totalenergy/data/annual/pdf/aer.pdf (accessed October 3, 2012)

FIGURE 5.9

Net summary capacity of operable nuclear units, 1957–2011

SOURCE: "Figure 9.2. Nuclear Power Plant Operations: Net Summer Capacity of Operable Units, 1957–2011," in *Annual Energy Review 2011*, U.S. Energy Information Administration, September 27, 2012, http://www.eia.gov/totalenergy/data/annual/pdf/aer.pdf (accessed October 3, 2012)

FIGURE 5.10

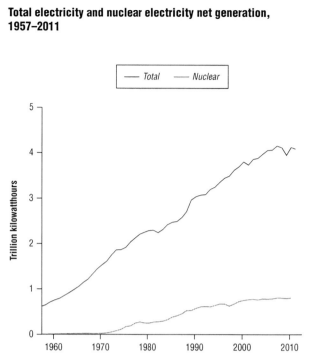

Total electricity and nuclear electricity net generation, 1957–2011

SOURCE: "Figure 9.2. Nuclear Power Plant Operations: Total Electricity and Nuclear Electricity Net Generation, 1957–2011," in *Annual Energy Review 2011*, U.S. Energy Information Administration, September 27, 2012, http://www.eia.gov/totalenergy/data/annual/pdf/aer.pdf (accessed October 3, 2012)

France produced 78% of its electricity from nuclear power, the largest percentage of any nation. Belgium and Slovakia were second, with 54% each, followed by Ukraine, at 47%, and Hungary, at 43%. The United States produced 19% of its electrical power using nuclear energy in 2011. Sixty-two additional nuclear power units were under construction or on order worldwide in 2011. (See Table 5.1.) China was expected to install the most new capacity (26,620 net megawatts), followed by Russia (9,297 net megawatts) and India (4,824 net megawatts).

The EIA forecasts in *Annual Energy Outlook 2012* (June 2012, http://www.eia.gov/forecasts/aeo/pdf/0383 (2012).pdf) long-term growth in nuclear power generation for countries that are and are not members of the Organisation for Economic Co-operation and Development (OECD). The OECD is a collection of dozens of mostly western nations that are devoted to global economic development. As shown in Figure 5.12, OECD members are expected to make small gains in nuclear power capacity by 2035. However, nonmembers, including Russia, China, and India, are projected to make large capacity additions.

NUCLEAR SAFETY ISSUES

Safety is a concern in all energy industries, but the nuclear power industry has unique safety concerns. The consequences of a destructive accident or terrorist attack at a nuclear power plant could be quite calamitous due to radiation release. One troubling scenario involves a meltdown of the reactor core following the failure of the cooling system. In this event, all the control rods holding the uranium fuel get so hot they melt, releasing their radioactive contents. As shown in Figure 5.2, a typical U.S. reactor is encased in a containment structure with walls that are made of concrete and steel several feet thick. Thus, a partial or even complete meltdown does not necessarily mean that radiation will escape from the reactor. It depends on the structural integrity of the containment walls.

Nuclear power has been generated commercially only since the late 1950s and early 1960s. However, during that time three major incidents have occurred in three different countries that have severely impacted the industry's safety reputation.

Three Mile Island

On March 28, 1979, the Three Mile Island nuclear facility near Harrisburg, Pennsylvania, was the site of the worst nuclear accident in U.S. history when one of its reactors overheated and suffered a partial meltdown. The emergency system was designed to dump water on the hot core of the reactor and spray water into the reactor building to stop the production of steam. During the accident, however, the valves leading to the emergency water pumps closed. Another valve was stuck in the open position, drawing water away from the core, which then became partially uncovered and began to melt.

TABLE 5.1

World nuclear power units, capacity, production, and shutdowns, by nation, 2011

Country	In operation		Under construction or on order		Nuclear power production GWh*	Shutdown
	Number of units	Capacity net MWe	Number of units	Capacity net MWe		
Argentina	2	935	1	692	5,894	0
Armenia	1	375	0	0	2,357	1[P]
Belgium	7	5,927	0	0	45,942	1[P]
Brazil	2	1,884	1	1,245	14,795	0
Bulgaria	2	1,906	2	1,906	15,264	4[P]
Canada	18	12,604	0	0	88,318	3[P]&4[L]
China	16	11,816	26	26,620	82,569	0
Czech Republic	6	3,678	0	0	26,696	0
Finland	4	2,716	1	1,600	22,266	0
France	58	63,130	1	1,600	423,509	12[P]
Germany	9	12,068	0	0	102,311	27[P]
Hungary	4	1,889	0	0	14,707	0
India	20	4,391	7	4,824	28,948	0
Iran	1	915	1	915	98	0
Italy	0	0	0	0	0	4[P]
Japan	50	44,215	2	2,650	156,182	9[P]&1[L]
Kazakhstan	0	0	0	0	0	1[P]
Rep. Korea	23	20,671	3	3,640	147,763	0
Lithuania	0	0	0	0	0	2[P]
Mexico	2	1,300	0	0	9,313	0
Netherlands	1	482	0	0	3,917	1[P]
Pakistan	3	725	2	630	3,843	0
Romania	2	1,300	0	0	10,811	0
Russia	33	23,643	11	9,297	162,018	5[P]
Slovakia	4	1,816	2	782	14,342	3[P]
Slovenia	1	688	0	0	5,902	0
South Africa	2	1,830	0	0	12,939	0
Spain	8	7,567	0	0	55,121	2[P]
Sweden	10	9,331	0	0	58,098	3[P]
Switzerland	5	3,263	0	0	25,694	1[P]
Ukraine	15	13,107	2	1,900	84,894	4[P]
United Kingdom	17	9,736	0	0	62,658	28[P]
United States	104	101,465	1	1,165	790,439	28[P]
Total	**436**	**370,499**	**62**	**59,245**	**2,517,980**	**139[P] & 5[L]**

GWh = gigawatt hours. MWe = megawatts electric.
[P]Permanent shutdown.
[L]Long-term shutdown.
*Annual electric power production from 2011.

SOURCE: Ivonne Couret et al., "Appendix S. Nuclear Power Units by Nation," in *U.S. Nuclear Regulatory Commission Information Digest 2012–2013*, vol. 24, U.S. Nuclear Regulatory Commission, Office of Public Affairs, August 2012, http://www.nrc.gov/reading-rm/doc-collections/nuregs/staff/sr1350/v24/sr1350v24.pdf (accessed October 14, 2012).

The accident did not result in any deaths or injuries to plant workers or to people in the nearby community. On average, area residents were exposed to less radiation than that of a chest X-ray. Nevertheless, the incident raised concerns about nuclear safety, which resulted in more rigorous safety standards in the nuclear power industry. Antinuclear sentiment was fueled as well, heightening Americans' wariness of nuclear power as an energy source.

The damaged nuclear reactor at Three Mile Island was permanently shut down after it underwent cleanup. However, an undamaged reactor at the plant continued to operate as of December 2012.

Chernobyl

On April 26, 1986, the most serious nuclear accident in history occurred at Chernobyl, a nuclear plant in what is now Ukraine (then part of the Soviet Union). At least 31 people died and hundreds were injured when one of the four reactors exploded during a badly run test. Millions of people were exposed to some levels of radiation when radioactive particles were released into the atmosphere. About 350,000 people were eventually evacuated from the area.

The cleanup was a huge project. Helicopters dropped tons of limestone, sand, clay, lead, and boron on the smoldering reactor to stop the radiation leakage and reduce the heat. Meanwhile, workers built a giant steel and cement sarcophagus to entomb the remains of the reactor and contain the radioactive waste.

The International Atomic Energy Agency indicates in *Chernobyl's Legacy: Health, Environmental and Socio-economic Impacts* (April 2006, http://www.iaea.org/Publications/Booklets/Chernobyl/chernobyl.pdf) that about 1,000 people involved in the initial cleanup,

FIGURE 5.11

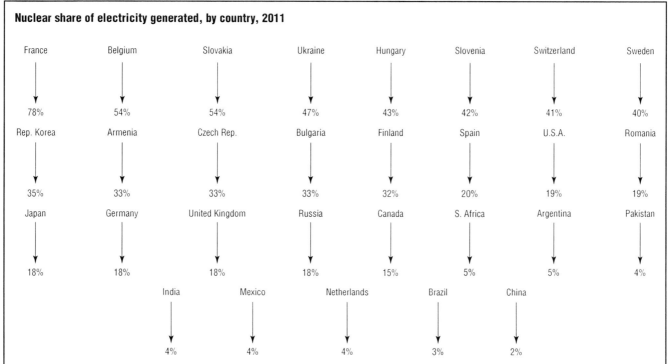

Nuclear share of electricity generated, by country, 2011

France	Belgium	Slovakia	Ukraine	Hungary	Slovenia	Switzerland	Sweden
78%	54%	54%	47%	43%	42%	41%	40%
Rep. Korea	Armenia	Czech Rep.	Bulgaria	Finland	Spain	U.S.A.	Romania
35%	33%	33%	33%	32%	20%	19%	19%
Japan	Germany	United Kingdom	Russia	Canada	S. Africa	Argentina	Pakistan
18%	18%	18%	18%	15%	5%	5%	4%
	India	Mexico	Netherlands	Brazil	China		
	4%	4%	4%	3%	2%		

Note: The country's short-form name is used.

SOURCE: Adapted from Ivonne Couret et al., "Figure 11. Nuclear Share of Electricity Generated by Country, 2011," in *U.S. Nuclear Regulatory Commission Information Digest 2012–2013*, vol. 24, U.S. Nuclear Regulatory Commission, Office of Public Affairs, August 2012, http://www.nrc.gov/reading-rm/doc-collections/nuregs/staff/sr1350/v24/sr1350v24.pdf (accessed October 14, 2012)

including emergency workers and the military, received high doses of radiation. Eventually, over 600,000 people were involved in decontamination and containment activities. The long-term effects of whatever exposure they received are being monitored.

Even though some of the evacuated land in Chernobyl has been declared fit for habitation again, several areas that received heavy concentrations of radiation are expected to be closed for decades.

Fukushima Daiichi

On March 11, 2011, an underwater earthquake triggered a tsunami that flooded the northeastern coast of Japan and killed nearly 20,000 people. The Fukushima Daiichi nuclear power plant is located on the northeastern coast approximately 150 miles (240 km) north of Tokyo. Three of the plant's six reactors suffered meltdowns after their cooling systems became inoperable following the earthquake and tsunami. Over subsequent days the reactors were plagued by partial and complete meltdowns and small hydrogen gas explosions. Some radioactive gases were released into the atmosphere from the damaged containment structures. The overheated reactors were finally cooled with seawater to prevent further releases.

Kevin Krolicki reports in "Fukushima Radiation Higher than First Estimated" (Reuters, May 24, 2012)

that more than a year after the accident occurred officials estimated that "the amount of radiation released in the first three weeks of the accident [was] about one-sixth the radiation released during the 1986 Chernobyl disaster." The Fukushima incident dampened the enthusiasm that had been building around the world for greater reliance on nuclear power. In *Annual Energy Outlook 2012*, the EIA notes "in the aftermath, governments in several countries that previously had planned to expand nuclear capacity—including Japan, Germany, Switzerland, and Italy—reversed course." In fact, the agency notes that Germany decided to phase out all of its nuclear power by 2025. In addition, the EIA predicts that nuclear power capacity in Japan will defy the trend for most other OECD countries and decline between 2010 and 2035. (See Figure 5.12.)

NUCLEAR WASTE ISSUES

Another safety (and environmental) issue related to nuclear power production is the creation of radioactive waste. Radioactive waste is produced at all stages of the nuclear fuel cycle, from the initial mining of the uranium to the final disposal of the spent fuel from the reactor. The term *radioactive waste* encompasses a broad range of material with widely varying characteristics. Some is barely radioactive and safe to handle, whereas other types are intensely hot and highly radioactive. Some waste

FIGURE 5.12

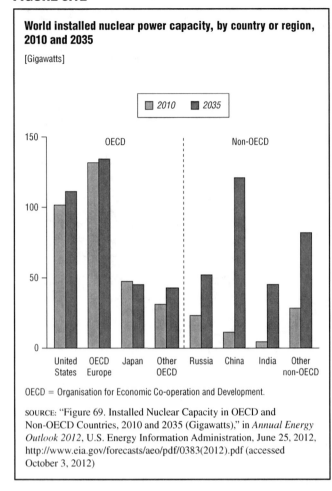

World installed nuclear power capacity, by country or region, 2010 and 2035

[Gigawatts]

OECD = Organisation for Economic Co-operation and Development.

SOURCE: "Figure 69. Installed Nuclear Capacity in OECD and Non-OECD Countries, 2010 and 2035 (Gigawatts)," in *Annual Energy Outlook 2012*, U.S. Energy Information Administration, June 25, 2012, http://www.eia.gov/forecasts/aeo/pdf/0383(2012).pdf (accessed October 3, 2012)

decays to safe levels of radioactivity in a matter of days or weeks, whereas other types will remain dangerous for thousands of years. The DOE and the NRC have defined the major types of radioactive waste that are associated with nuclear power generation in the United States.

Disposing of radioactive waste is unquestionably one of the major problems associated with the development of nuclear power. The highly toxic wastes must be isolated from the environment until the radioactivity decays to a safe level. That time period can last from several years to several millennia, depending on the radioactivity of the waste. Even though U.S. policy is based on the assumption that radioactive waste can be disposed of safely, new storage and disposal facilities for all types of radioactive waste have frequently been delayed or blocked by concerns about safety, health, and the environment.

Waste Types and Disposal

URANIUM MILL TAILINGS. Uranium mill tailings are sandlike wastes produced in uranium refining operations. Even though they emit low levels of radiation, their large volume poses a hazard, particularly from radon emissions and groundwater contamination. Mill tailings are usually deposited in large piles next to the mill that processed the ore and must be covered to prevent radiation problems.

LOW-LEVEL WASTE. Low-level waste (LLW) includes trash (such as wiping rags, swabs, and syringes), contaminated clothing (such as shoe covers and protective gloves), and hardware (such as luminous dials, filters, and tools). This waste comes from nuclear reactors, industrial users, government users (but not nuclear weapons sites), research universities, and medical facilities. In general, LLW decays relatively quickly (in 10 to 100 years). According to Couret et al., as of 2012 four licensed LLW facilities were operating in the United States: in Barnwell, South Carolina; in Hanford, Washington; in Clive, Utah; and in Andrews, Texas.

The Low-Level Radioactive Waste Policy Amendments Act of 1985 encouraged states to enter into compacts, which are legal agreements among states for low-level radioactive waste disposal. Each compact is responsible for the development of disposal capacity for the LLW generated within the compact; however, as of December 2012 new disposal sites had yet to be built. Nuclear power facilities that are located in compacts without existing LLW disposal sites must petition the compact to export their low-level radioactive waste to one of the four operating LLW disposal sites.

SPENT NUCLEAR FUEL. Spent nuclear fuel is fuel that has exhausted its usefulness in a nuclear power reactor. It is called high-level waste because it is highly radioactive and requires very long containment times, typically thousands of years.

For decades the United States has pursued development of a national repository for its high-level wastes; however, finding a suitable site acceptable to all parties involved has proven impossible. In 1987 the federal government began investigating Yucca Mountain, a barren wind-swept mountain in the desert about 100 miles (161 km) northwest of Las Vegas, Nevada. The project was highly controversial from the start and has been fought bitterly by the state of Nevada and environmentalists. After years of court challenges and political wrangling the administration of President Barack Obama (1961–) dropped the site in 2010 as a repository candidate. Obama refused to provide further federal funding toward licensing Yucca Mountain and established the Blue Ribbon Commission on America's Nuclear Future (BRC) to recommend alternative options for consideration. In January 2012 the BRC issued its final report, *Report to the Secretary of Energy* (http://brc.gov/sites/default/files/documents/brc_finalreport_jan2012.pdf), in which it noted "this nation's failure to come to grips with the nuclear waste issue has already proved damaging and costly and it will be more damaging and more costly the longer it continues." The BRC recommended numerous legislative and policy changes as first steps toward finding a permanent solution to the high-level waste disposal problem.

Despite the Obama administration's rejection of the Yucca Mountain site, some proponents of the site have continued to lobby on its behalf. In "Election Hasn't Changed Yucca Mountain's Future" (*Las Vegas Review-Journal*, November 11, 2012), Steve Tetreault reports that a court ruling was expected before year-end 2012 on a lawsuit that was filed by parties trying to force the federal government to resume licensing the project. According to Tetreault, the parties include local officials from Nye County, the county in which Yucca Mountain is located. As of December 2012, a final ruling had not been issued in the case.

Thus, as of December 2012 the United States had no central repository for spent nuclear fuel. The fuel continued to be stored either at the nuclear power plants that generated it or at other secure locations. Figure 5.13 shows the states storing spent nuclear fuel as of year-end 2011. The five states with the largest stores were Illinois, Pennsylvania, South Carolina, New York, and North Carolina. The U.S. Government Accountability Office (GAO) notes in *Spent Nuclear Fuel: Accumulating Quantities at Commercial Reactors Present Storage and*

Other Challenges (August 2012, http://www.gao.gov/assets/600/593745.pdf) that approximately 77,200 tons (70,000 t) of spent fuel were in storage as of August 2012. Most (74%) of the waste was being stored in water pools. The remaining 26% had cooled enough to be transferred into dry storage casks. (See Figure 5.14.) The GAO estimates that the amount of spent fuel will grow at a rate of approximately 2,200 tons (2,000 t) per year and may grow to 154,000 tons (140,000 t) or more in total because a permanent storage or disposal facility is likely decades away from development.

One method used by other countries to reduce spent nuclear fuel volumes is reprocessing. In "Reprocessing" (November 8, 2012, https://www.nrc.gov/materials/reprocessing.html), the NRC explains that reprocessing involves separating spent nuclear fuel into two streams: still viable fuel that can be reused and waste for disposal. The nuclear fuel cycle shown in Figure 5.3 includes a reprocessing facility; however, as of December 2012 no such facility existed in the United States. The nation has long rejected reprocessing, primarily because the technique separates plutonium from uranium. Plutonium can be used to build nuclear bombs. Presently, the plutonium within spent nuclear fuel is dispersed among large amounts of uranium and other "contaminants." Thus, the plutonium is not a

FIGURE 5.13

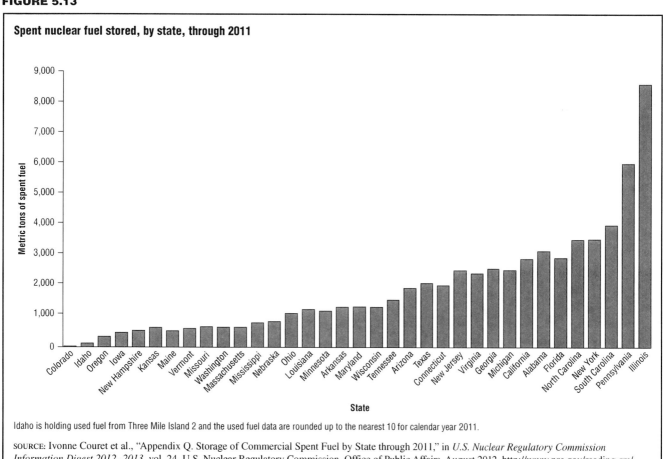

Spent nuclear fuel stored, by state, through 2011

Idaho is holding used fuel from Three Mile Island 2 and the used fuel data are rounded up to the nearest 10 for calendar year 2011.

SOURCE: Ivonne Couret et al., "Appendix Q. Storage of Commercial Spent Fuel by State through 2011," in *U.S. Nuclear Regulatory Commission Information Digest 2012–2013*, vol. 24, U.S. Nuclear Regulatory Commission, Office of Public Affairs, August 2012, http://www.nrc.gov/reading-rm/doc-collections/nuregs/staff/sr1350/v24/sr1350v24.pdf (accessed October 14, 2012)

FIGURE 5.14

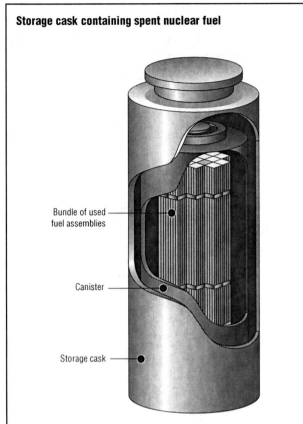

Storage cask containing spent nuclear fuel

Bundle of used
fuel assemblies

Canister

Storage cask

SOURCE: Adapted from Ivonne Couret et al., "Spent Fuel Storage Cask," in *U.S. Nuclear Regulatory Commission Information Digest 2012–2013*, vol. 24, U.S. Nuclear Regulatory Commission, Office of Public Affairs, August 2012, http://www.nrc.gov/reading-rm/doc-collections/nuregs/staff/sr1350/v24/sr1350v24.pdf (accessed October 14, 2012)

tempting target for terrorists or other parties seeking weapons-grade plutonium. Couret et al. note that the NRC "has not received an application for a reprocessing facility" in the United States. However, they indicate that the commission has a regulatory framework in place in case it does receive such an application.

NUCLEAR POWER PRICES

As noted earlier, the commercial nuclear power industry is devoted entirely to electricity production. In the United States electricity is also produced by the combustion of fossil fuels, mainly coal and natural gas. Because electricity producers feed their electricity to the nation's electrical grid for consumption by consumers, it is difficult to distinguish between the prices for electricity produced from various fuel sources. However, there are specific supply and demand factors that affect nuclear power pricing.

Analysts agree that one cost differential between nuclear-based electricity and electricity from other sources is capital costs. Construction of a nuclear power plant is extremely expensive. However, most of the United

States' nuclear power plants were built decades ago. Couret et al. provide the following age breakdown for the nation's 104 operable units at year-end 2012:

- Two reactors—10 to 19 years in operation
- 37 reactors—20 to 29 years in operation
- 50 reactors—30 to 39 years in operation
- 15 reactors—40 years or more in operation

Thus, the high capital costs that are associated with building the reactors have long been absorbed by the industry and are not a significant factor in current pricing. The same holds true for initial licensing costs and other costs that are incurred before plants begin operating.

The Nuclear Energy Institute (NEI) is a lobbying group for the nuclear power industry. In "Fuel as a Percentage of Electric Power Production Costs: 2011" (May 2012, http://www.nei.org/filefolder/Fuel_as_Percent_Electric_Production_Costs.ppt), the NEI estimates that operation and maintenance costs account for 69% of the total cost of nuclear power production, while fuel costs account for 31% of the total. Fuel costs are broken down as follows: uranium purchases (42%), uranium processing (31%), the Nuclear Waste Fund (15%), fabrication (8%), and conversion (4%). The Nuclear Waste Fund is administered by the federal government. According to the DOE, in "2011 Secretarial Determination of the Adequacy of the Nuclear Waste Fund Fee" (December 2011, http://www.ocrwm.doe.gov/sites/prod/files/2011%20Secretarial%20Fee%20Adequacy%20Determination.PDF), the fund was established by law in 1983 and requires civilian nuclear power producers to pay a fee of one-tenth cent per kilowatt-hour of electricity generated. The DOE notes that the fund collects around $750 million in fees and earns approximately $1 billion in interest each year. As of December 2011, the fund totaled nearly $17 billion.

The Nuclear Waste Fund represents an expense that is imposed by the government on the nuclear power industry; however, the government has also taken actions to financially support and promote the industry. These actions help lower the production price of electricity that is generated by nuclear power.

In *Direct Federal Financial Interventions and Subsidies in Energy in Fiscal Year 2010* (July 2011, http://www.eia.gov/analysis/requests/subsidy/pdf/subsidy.pdf), the EIA notes that in 2010 the nuclear power industry received nearly $2.5 billion in tax breaks, research and development funds, and other types of financial assistance from the federal government. Another $971 million was devoted to general electricity distribution and transmission. Together, these funds amounted to around 9% of the $37.2 billion in total energy-specific subsidies and support provided that year.

The nuclear power industry has also benefitted from liability limits that have been set by federal law. According to the EIA, the law requires each nuclear power plant operator to obtain liability insurance. As of 2011, the insurance covered a maximum of around $400 million. Any damages exceeding that amount are to be paid by collecting funds from all other companies owning commercial nuclear reactors. However, the EIA indicates that the law limits the total liability of all the owners to around $12 billion. Thus, any costs of a nuclear accident in excess of $12 billion might have to be borne by U.S. taxpayers.

OUTLOOK FOR DOMESTIC NUCLEAR ENERGY

The EIA indicates in *Annual Energy Review 2011* that between 2007 and 2011 the NRC granted four Early Site Permits for nuclear power plants. Also, between 2007 and 2011 the NRC received 18 Combined License applications. According to the NRC (March 29, 2012, http://www.nrc.gov/reactors/new-reactors/col.html), a Combined License grants a recipient permission to construct and operate a nuclear power plant at a specific site. An NRC map (http://www.nrc.gov/reactors/new-reactors/col/new-reactor-map.html) shows the locations of projected new nuclear power reactors as of March 2012. All the sites were in the eastern United States. In February 2012 the NRC announced that it had approved a Combined License permit for two new units at an existing nuclear power plant in Georgia. These are the first new units permitted since the late 1970s. The Southern Company, the plant's owner, notes in "The Plan" (2012, http://www.southerncompany.com/nuclearenergy/plan.aspx) that the new units at the Alvin W. Vogtle Electric Generation Plant near Augusta, Georgia, are expected to begin operating by 2016 or 2017.

In *Annual Energy Outlook 2012*, the EIA predicts energy production in the future using 2010 levels as a baseline. The agency expects that the total capacity of U.S. nuclear power plants will increase from 101 gigawatts in 2010 to 111 gigawatts in 2035. About 8.5 gigawatts of capacity will come from new plants and 7.6 gigawatts from uprates (expansions at existing plants). However, these gains will be somewhat offset by capacity losses from retiring older nuclear power plants. Overall, the EIA expects that electricity generation from nuclear power plants will grow by 10% between 2010 and 2035. However, the share that nuclear power contributes to electricity generation in the United States is expected to decrease slightly to 18% in 2035. More information about projected future electricity supply and demand is included in Chapter 8.

CHAPTER 6
RENEWABLE ENERGY

Renewable energy is energy that is naturally regenerated and virtually unlimited. Before the 18th century, most energy came from renewable sources. People burned wood for heat, used sails to harness the wind and propel boats, and installed water wheels on streams to run mills to grind grain. The large-scale shift to fossil fuels, such as coal, began with the Industrial Revolution (1760–1848), a period that was marked by the rise of factories, first in Europe and then in North America.

Until the early 1970s most Americans were perfectly content to rely on fossil fuels for their energy needs. However, an embargo on imported oil ended this carefree approach. Throughout the United States people waited in line to fill their gas tanks—in some places gasoline was rationed—and lower heat settings for offices and homes were encouraged. In a country where mobility and convenience were highly valued, the 1970s oil crisis was a shock to the system. Developing alternative sources of energy to supplement and perhaps eventually replace fossil fuels suddenly became important. Over the following decades financial incentives and technological advancements helped alternative energy providers build their industries.

Renewable energy sources are relative newcomers to the energy marketplace and, in many cases, must compete directly with fossil fuels for consumers. This is a difficult task. The systems and infrastructure for fossil fuel production and delivery have long been in place, and Americans are accustomed to them. As noted in previous chapters, fossil fuels are forecast to maintain their prominent role as energy sources for decades to come. Even so, renewable sources have carved out a niche in the market, and that niche seems destined to grow.

UNDERSTANDING RENEWABLE ENERGY

Like fossil fuels, renewable energy sources are produced naturally, but unlike fossil fuels, renewable energy sources do not take millennia to form. Some are available immediately and regenerate constantly. Examples include wind, sunlight, tides, and the heat stored beneath the earth's crust. In addition, water bodies can serve as renewable energy sources when the water is used as a working fluid to turn turbines and produce electricity.

Wood and other plant-derived materials are considered to be renewable energy sources because trees and crops can be planted as needed and harvested in relatively short time periods (i.e., months to decades as opposed to millennia). Likewise, combustible wastes produced by humans (e.g., household garbage and used tires) are considered to be renewable energy sources. Biomass is a generic term for various biologically based materials that can be used for energy production. Biomass includes wood, sawdust, organic waste (e.g., food and paper scraps), agricultural by-products (e.g., corn husks), liquid fuels derived from plants (e.g., corn-based ethanol), and gases such as methane that naturally emanate from landfills in which household garbage is undergoing organic degradation.

As noted in Chapter 1, energy consumption involves conversion from one type of energy to another type. Many of the renewable sources are converted from their original forms to electrical energy. Electricity is a high-demand product, and once generated (by whatever means) it can be fed into the existing electrical grid that serves the nation. Some renewable sources serve additional markets. For example, biomass sources such as landfill methane and corn-based ethanol can be burned, just like fossil fuels, to heat buildings and power vehicles and industrial equipment. Whatever their intended end-uses may be, renewables must overcome logistical, technological, and economical challenges to get to consumers.

It is also important to understand that even though renewables begin as natural sources, they may be heavily processed before they actually provide energy to consumers.

Industrial processing requires energy—that is, it has an energy price. For example, chipping wood into small pieces so it is easier to burn is energy intensive. Likewise, converting crops, such as corn, into liquid fuel is a highly industrial process. Because U.S. industry is heavily dependent on fossil fuels, the reality is that fossil fuels are burned during the processing of many renewable sources. This energy price has to be considered when evaluating the relative advantages and disadvantages of alternative fuels.

MARKET FACTORS

The renewables market is driven by concerns about energy production from nuclear energy and fossil fuels. One major concern relates to environmental impacts. Nuclear power generates radioactive waste that can be hazardous for thousands of years. The combustion of fossil fuels releases emissions that degrade air quality and contribute to global warming and climate change. Renewable fuels do not generate radioactive waste. The noncombustible renewables (solar, wind, tides, hydropower, and geothermal) do not emit the air pollutants that are associated with fossil fuels. The combustible renewables, by virtue of their organic makeup, do contribute such emissions, but to a lesser extent than fossil fuels. Energy self-sufficiency is another concern that drives interest in renewable fuels because they can be produced domestically.

Although environmental impacts and self-sufficiency are powerful motivators for developing alternative fuels, both concerns pale in comparison to economic considerations. Many renewable energy source technologies are relatively immature and still in the research and development (R&D) stage. As such, they are more expensive to produce commercially than fossil fuels. When fossil fuels drop in price, interest wanes in the business world for developing alternative fuels because there is less profit to be made. In addition, the impetus for renewables development declines when "new" domestic sources of fossil fuels become available. As described in Chapters 2 and 3, domestic production of oil and natural gas from unconventional geological sources, such as shale and tar sands, is expected to skyrocket in coming decades. These sources will likely help keep domestic fossil fuel prices low, further damaging the economic competitiveness of renewables.

Another stumbling block for renewables in the short term is that many of them are devoted to electricity generation. The domestic electric power sector has a glut of existing natural gas capacity in the United States. In *Annual Energy Outlook 2012* (June 2012, http://www.eia.gov/forecasts/aeo/pdf/0383(2012).pdf), the Energy Information Administration (EIA) within the U.S. Department of Energy (DOE) notes that there is "relatively little projected need for new generation capacity of any type, including renewables, for the remainder of the current decade, primarily because there is an abundance of existing natural gas fired capacity that can be operated at higher capacity factors." The agency does not expect demand for renewables-based generating capacity to grow until after 2020.

Federal Government Support

When interest in developing renewable energy sources became keen during the 1970s, the government began providing financial support to the fledgling industry. As described in Chapter 1, the federal government uses a variety of incentives to encourage domestic energy development. The Congressional Budget Office (CBO) notes in *Federal Financial Support for the Development and Production of Fuels and Energy Technologies* (March 2012, http://www.cbo.gov/sites/default/files/cbofiles/attachments/03-06-FuelsandEnergy_Brief.pdf) that in 2011 tax preferences for renewable energy totaled $12.9 billion, or 63% of the $20.5 billion total. In addition, the CBO indicates that the DOE spent approximately $4 billion on subsidies to the energy industry between 2009 and 2012 and that those subsidies "primarily" went to renewable energy producers or supporting industries. The CBO notes that most of the renewable energy provisions expired at the end of 2011, meaning that they were only temporary measures. By contrast, the provisions supporting fossil fuels and nuclear energy did not have expiration dates.

According to the CBO, government support via tax breaks and subsidies for the fossil fuel industry dates back to 1916. Provisions benefitting the renewables industry began during the 1970s but remained relatively small for three decades. The CBO indicates that tax incentives for fossil fuels typically accounted for more than two-thirds of total energy tax incentives through 2007. Legislative changes, particularly the Energy Policy Act of 2005, helped turn the focus toward renewables, which have since garnered a larger percentage of the financial support given to energy overall.

Nevertheless, critics complain that the incentives are miniscule in comparison with the many billions of dollars that have been pumped into the fossil fuel industry for nearly a century. According to the CBO, government support is particularly important during the initial R&D stages of an industry. In the second decade of the 21st century the fossil fuel industry is extremely mature, whereas the renewables industry is still largely doing initial R&D tasks. (It should be noted that this is not true for the hydroelectric sector, which has operated for many decades.) Some analysts believe the government should provide more funding support for renewables at this crucial time in their development and make their tax preferences permanent to boost the competitiveness of the industry.

The federal government also supports renewables development through mandates and standards. For example, the Energy Independence and Security Act of 2007 includes a Renewable Fuels Standard that mandates the use of specified amounts of biofuels by 2022. This standard will be explained in more detail later in this chapter.

State Governments Support

The EIA indicates in *Annual Energy Outlook 2012* that as of June 2012, 30 states and the District of Columbia had mandatory (enforceable) renewable portfolio standards or similar laws. The agency notes that "under such standards, each State determines its own levels of renewable generation, eligible technologies, and noncompliance penalties." For example, California requires that renewables account for 33% of electricity sales in the state by 2020. According to the EIA, by 2015 all the state standards together will account for around 7% of the total projected U.S. electricity sales. By 2020 the percentage will be closer to 10%. In both forecasts the agency assumes that the standards will be fully implemented and binding.

PRODUCTION AND CONSUMPTION

In *Annual Energy Review 2011* (September 2012, http://www.eia.gov/totalenergy/data/annual/pdf/aer.pdf), the EIA indicates that in 2011 the United States consumed 9.1 quadrillion British thermal units (Btu) of renewable energy, or 9% of the nation's total energy consumption. Hydroelectric power was the largest source, providing 3.2 quadrillion Btu, or 35% of the total renewable energy consumption. (See Figure 6.1.) Wood was second at 2 quadrillion Btu (22% of the total). Biofuels were the third-largest source, with 1.9 quadrillion Btu (21% of the total). Smaller sources included wind (1.2 quadrillion Btu, or 13% of the total), waste (477 trillion Btu, or 5% of the total), geothermal (226 trillion Btu, or 2% of the total), and solar/photovoltaic energy (158 trillion Btu, or 2% of the total).

Figure 6.2 shows historical consumption of renewables between 1949 and 2011. Wood and hydroelectric power were nearly equally consumed during the late 1940s and early 1950s. Hydroelectric power use then surged above that of wood and became the dominant renewable source. Other sources, including biofuels and wind, came into use much later in the 20th century. Consumption of these sources remained relatively low until the first decade of the 21st century, when it increased dramatically. In 2011 biofuels and wood were nearly equally consumed, and biofuels seemed poised to surpass wood as the combustible renewable of choice.

Figure 6.3 provides a breakdown of renewables consumption in 2011 by end-use sectors and the electric power sector. The electric power sector accounted for

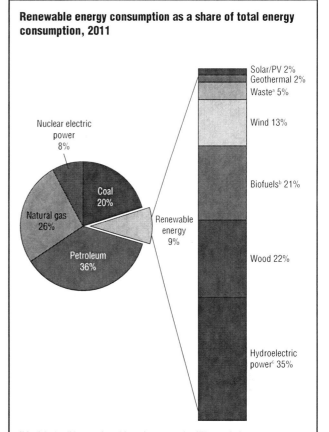

FIGURE 6.1

Renewable energy consumption as a share of total energy consumption, 2011

[a]Municipal solid waste from biogenic sources, landfill gas, sludge waste, agricultural byproducts, and other biomass.
[b]Fuel ethanol (minus denaturant) and biodiesel consumption, plus losses and co-products from the production of fuel ethanol and biodiesel.
[c]Conventional hydroelectric power.
Notes: Sum of components may not equal 100 percent due to independent rounding.

SOURCE: Figure 10.1. Renewable Energy Consumption by Major Source: Renewable Energy As Share of Total Primary Energy Consumption, 2011," in *Annual Energy Review 2011*, U.S. Energy Information Administration, September 27, 2012, http://www.eia.gov/totalenergy/data/annual/pdf/aer.pdf (accessed October 3, 2012)

4.9 quadrillion Btu, which was more than half (54%) of the total consumption. The industrial and transportation sectors were the next biggest consumers of renewable energy sources in 2011. The residential and commercial sectors were much smaller users.

HYDROPOWER

Hydropower has been used for decades in the United States to generate electricity. Hydropower facilities convert the energy of flowing water into mechanical energy, turning turbines to create electricity. In "What Is the Role of Hydroelectric Power in the United States?" (August 29, 2012, http://www.eia.gov/energy_in_brief/hydropower.cfm), the EIA divides hydropower into two broad categories: conventional and nonconventional.

Conventional hydropower plants use either flowing water (a run-of-river plant) or water backed up behind a

FIGURE 6.2

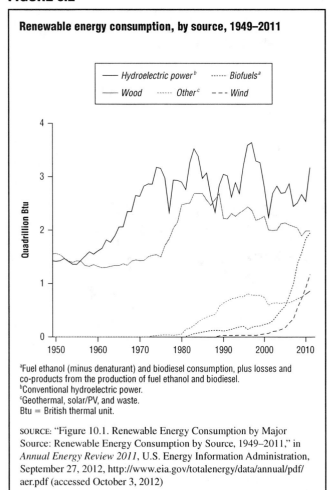

Renewable energy consumption, by source, 1949–2011

[legend]
Hydroelectric power[b] ...... Biofuels[a]
Wood ...... Other[c] --- Wind

[a]Fuel ethanol (minus denaturant) and biodiesel consumption, plus losses and co-products from the production of fuel ethanol and biodiesel.
[b]Conventional hydroelectric power.
[c]Geothermal, solar/PV, and waste.
Btu = British thermal unit.

SOURCE: "Figure 10.1. Renewable Energy Consumption by Major Source: Renewable Energy Consumption by Source, 1949–2011," in *Annual Energy Review 2011*, U.S. Energy Information Administration, September 27, 2012, http://www.eia.gov/totalenergy/data/annual/pdf/aer.pdf (accessed October 3, 2012)

FIGURE 6.3

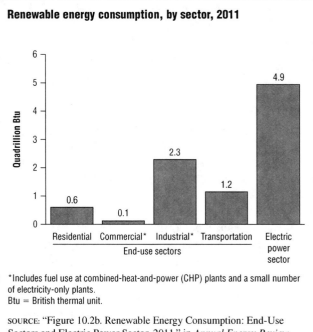

Renewable energy consumption, by sector, 2011

*Includes fuel use at combined-heat-and-power (CHP) plants and a small number of electricity-only plants.
Btu = British thermal unit.

SOURCE: "Figure 10.2b. Renewable Energy Consumption: End-Use Sectors and Electric Power Sector, 2011," in *Annual Energy Review 2011*, U.S. Energy Information Administration, September 27, 2012, http://www.eia.gov/totalenergy/data/annual/pdf/aer.pdf (accessed October 3, 2012)

FIGURE 6.4

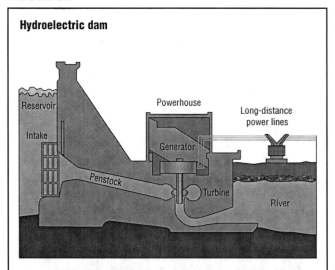

Hydroelectric dam

SOURCE: "Hydroelectric Dam," in *Hydropower Explained: Energy from Moving Water*, U.S. Energy Information Administration, March 21, 2012, http://www.eia.gov/energyexplained/index.cfm?page=hydropower_home (accessed October 15, 2012)

dam (a storage plant) to produce electricity. A run-of-river plant produces electricity as river flows allow. The flows can be quite variable and depend heavily on weather events, such as rainfall and snowmelt. By contrast, a storage plant relies on a dam to create an ever-ready reservoir of water that can be tapped as needed. (See Figure 6.4.) The dam creates a height from which water can flow at a fast rate. When the water reaches the power plant at the bottom of the dam, it pushes the turbine blades that are attached to the electrical generator. Whenever power is needed, the valves are opened, the moving water spins the turbines, and the generator produces electricity.

Even though there are various nonconventional hydropower means for generating electricity, the EIA notes that only one of these methods—pumped storage—was in "wide commercial use" as of 2012. Pumped-storage systems pump water from a low elevation to a high elevation within the plant during times of low demand for electricity. In essence, they reuse some of the water that has already been used to generate electricity. However, pumping is an energy-intensive process that can use more energy than is actually generated from the pumped water.

Figure 6.5 shows the locations of conventional and pumped-storage hydroelectric generators in and around the United States as of 2011. They are heavily concentrated in the river valleys of the central Atlantic states, the Pacific Northwest, and California. Most of the dams were built decades ago as part of massive federal programs that were designed to decrease flooding and provide fresh water

FIGURE 6.5

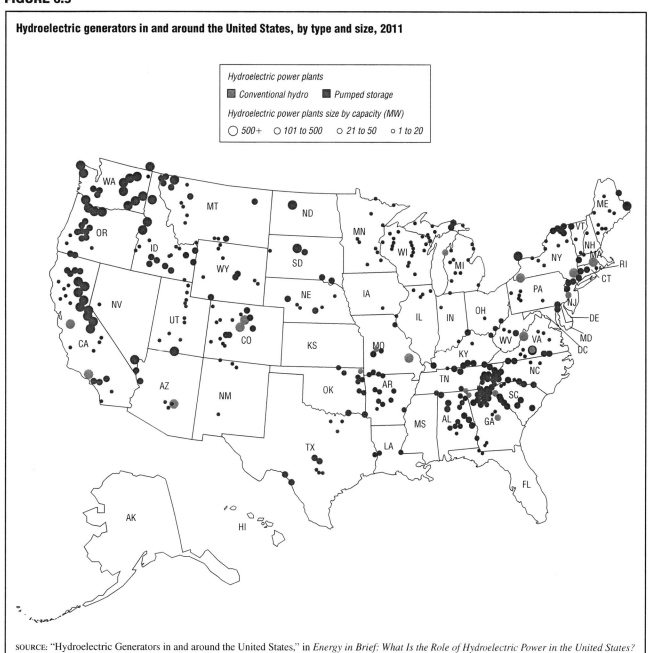

Hydroelectric generators in and around the United States, by type and size, 2011

Hydroelectric power plants
▣ Conventional hydro   ▪ Pumped storage
Hydroelectric power plants size by capacity (MW)
○ 500+   ○ 101 to 500   ○ 21 to 50   ○ 1 to 20

SOURCE: "Hydroelectric Generators in and around the United States," in *Energy in Brief: What Is the Role of Hydroelectric Power in the United States?* U.S. Energy Information Administration, August 29, 2012, http://www.eia.gov/energy_in_brief/images/charts/hydro_generators_map-large.jpg (accessed October 15, 2012)

supplies and electricity to the public. As a result, nearly all the best sites for large hydropower plants are already being used in the United States. Additionally, environmental groups strongly protest the construction of new dams, pointing to ruined streams, dried up waterfalls, and altered aquatic habitats.

Domestic Production and Consumption

According to the EIA, in *Annual Energy Review 2011*, hydroelectric power accounted for 8% of U.S. electricity production in 2011. Among renewable sources that are used for electricity production, hydropower is, by far, the dominant source. In 2011 it provided 3,153 trillion Btu, or 64% of the 4,945 trillion Btu consumed that year. The next closest competitor was wind power, which provided 1,168 trillion Btu, or 24% of the total.

The EIA notes that 4,105.7 billion kilowatt-hours (kWh) of net electricity generation occurred in 2011. Conventional hydroelectric power accounted for 325.1 billion kWh, or 8% of the total. The EIA indicates that hydroelectric generation from pumped storage was actually negative (−5.9 billion kWh) in 2011 because more energy was used for pumping the water than was garnered from production.

Domestic Market Factors

The hydroelectric power industry has been supported by the federal government for many decades. The CBO does not provide in *Federal Financial Support for the Development and Production of Fuels and Energy Technologies* a breakdown by source of the billions of dollars in federal incentives that the energy industry received in 2011. However, such a breakdown is available for 2010 from the EIA. In *Direct Federal Financial Interventions and Subsidies in Energy in Fiscal Year 2010* (July 2011, http://www.eia.gov/analysis/requests/subsidy/pdf/subsidy.pdf), the agency indicates that the hydropower industry received $216 million in direct expenditures, tax breaks, R&D funds, and other means of financial support in 2010. This was around 1.5% of the $14.7 billion that was allocated for the renewables industry as a whole. In addition, the hydropower industry benefitted from another $971 million that was devoted to enhancing electricity transmission and distribution in the nation.

The federal government not only supports the hydropower industry with financial incentives but is also an active participant in the industry. The Tennessee Valley Authority and the Power Marketing Administration are federally owned corporations that have long provided hydroelectricity to consumers in the south-central and western United States, respectively. According to the EIA, the federal government "brings to market large amounts of electricity," and that electricity is generated and priced for the public good. As a result, these government-sponsored endeavors help lower hydropower prices in the marketplace.

The Domestic Outlook

The EIA predicts in *Annual Energy Outlook 2012* domestic capacity and generation by renewable sources in 2035 using 2010 levels as a baseline. The agency expects conventional hydropower capacity (net summer) to grow by only 0.2% annually between 2010 and 2035. (See Table 6.1.) According to the EIA (2012, http://www.eia.gov/tools/glossary/index.cfm?id=net%20summer%20capacity), net summer capacity is the maximum output of the generating equipment during the peak summer demand time (between June 1 and September 30) less any capacity used by the generating station itself. Conventional hydropower generation by the electric power sector is predicted to increase by 0.8%, from 255.3 billion kWh in 2010 to 310.1 billion kWh in 2035. Conventional hydropower was a minor source of energy in 2010 to sectors other than the electric power sector and is expected to remain so in 2035.

As shown in Figure 6.6, the EIA forecasts that by around 2019 nonhydropower renewables (mainly wind, solar, and biomass) will begin generating more electricity than hydropower in the United States.

TABLE 6.1

Renewable energy generating capacity and generation, 2010 and 2035

[Gigawatts, unless otherwise noted]

Net summer capacity and generation	Reference case 2010	Reference case 2035	Annual growth 2010–2035 (percent)
Electric power sector[a]			
Net summer capacity			
Conventional hydropower	78.03	81.25	0.2%
Geothermal[b]	2.37	6.30	4.0%
Municipal waste[c]	3.30	3.36	0.1%
Wood and other biomass[d]	2.45	2.89	0.7%
Solar thermal	0.47	1.36	4.3%
Solar photovoltaic[e]	0.38	8.18	13.0%
Wind	39.05	66.65	2.2%
Offshore wind	0.00	0.20	—
Total electric power sector capacity	**126.06**	**170.19**	**1.2%**
Generation (billion kilowatthours)			
Conventional hydropower	255.32	310.08	0.8%
Geothermal[b]	15.67	46.54	4.5%
Biogenic municipal waste[f]	16.56	14.67	−0.5%
Wood and other biomass	11.51	49.28	6.0%
Dedicated plants	10.15	10.37	0.1%
Cofiring	1.36	38.92	14.4%
Solar thermal	0.82	2.86	5.1%
Solar photovoltaic[e]	0.46	20.19	16.4%
Wind	94.49	189.92	2.8%
Offshore wind	0.00	0.75	—
Total electric power sector generation	**394.82**	**634.30**	**1.9%**
End-use sectors[g]			
Net summer capacity			
Conventional hydropower[h]	0.33	0.33	0.0%
Geothermal	0.00	0.00	—
Municipal waste[i]	0.35	0.35	0.0%
Biomass	4.56	13.81	4.5%
Solar photovoltaic[e]	2.05	13.33	7.8%
Wind	0.36	2.74	8.5%
Total end-use sector capacity	**7.65**	**30.57**	**5.7%**
Generation (billion kilowatthours)			
Conventional hydropower[h]	1.76	1.75	−0.0%
Geothermal	0.00	0.00	—
Municipal waste[i]	2.02	2.79	1.3%
Biomass	26.10	96.17	5.4%
Solar photovoltaic[e]	3.21	20.91	7.8%
Wind	0.47	3.56	8.5%
Total end-use sector generation	**33.56**	**125.17**	**5.4%**
Total, all sectors			
Net summer capacity			
Conventional hydropower	78.36	81.58	0.2%
Geothermal	2.37	6.30	4.0%
Municipal waste	3.65	3.71	0.1%
Wood and other biomass[d]	7.00	16.71	3.5%
Solar[e]	2.90	22.87	8.6%
Wind	39.41	69.59	2.3%
Total capacity, all sectors	**133.70**	**200.76**	**1.6%**
Generation (billion kilowatthours)			
Conventional hydropower	257.08	311.83	0.8%
Geothermal	15.67	46.54	4.5%
Municipal waste	18.59	17.46	−0.3%
Wood and other biomass	37.61	145.45	5.6%
Solar[e]	4.48	43.96	9.6%
Wind	94.95	194.23	2.9%
Total generation, all sectors	**428.38**	**759.46**	**2.3%**

TABLE 6.1

Renewable energy generating capacity and generation, 2010 and 2035 [CONTINUED]

[Gigawatts, unless otherwise noted]

[a]Includes electricity-only and combined heat and power plants whose primary business is to sell electricity, or electricity and heat, to the public.
[b]Includes both hydrothermal resources (hot water and steam) and near-field enhanced geothermal systems (EGS). Near-field EGS potential occurs on known hydrothermal sites, however this potential requires the addition of external fluids for electricity generation and is only available after 2025.
[c]Includes municipal waste, landfill gas, and municipal sewage sludge. Incremental growth is assumed to be for landfill gas facilities.
All municipal waste is included, although a portion of the municipal waste stream contains petroleum-derived plastics and other non-renewable sources.
[d]Facilities co-firing biomass and coal are classified as coal.
[e]Does not include off-grid photovoltaics (PV). Based on annual PV shipments from 1989 through 2009, EIA estimates that as much as 245 megawatts of remote electricity generation PV applications (i.e., off-grid power systems) were in service in 2009, plus an additional 558 megawatts in communications, transportation, and assorted other non-grid-connected, specialized applications. The approach used to develop the estimate, based on shipment data, provides an upper estimate of the size of the PV stock, including both grid-based and off-grid PV. It will overestimate the size of the stock, because shipments include a substantial number of units that are exported, and each year some of the PV units installed earlier will be retired from service or abandoned.
[f]Includes biogenic municipal waste, landfill gas, and municipal sewage sludge. Incremental growth is assumed to be for landfill gas facilities. Only biogenic municipal waste is included. The U.S. Energy Information Administration estimates that in 2010 approximately billion kilowatthours of electricity were generated from a municipal waste stream containing petroleum-derived plastics and other non-renewable sources.
[g]Includes combined heat and power plants and electricity-only plants in the commercial and industrial sectors; and small on-site generating systems in the residential, commercial, and industrial sectors used primarily for own-use generation, but which may also sell some power to the grid.
[h]Represents own-use industrial hydroelectric power.
[i]Includes municipal waste, landfill gas, and municipal sewage sludge. All municipal waste is included, although a portion of the municipal waste stream contains petroleum-derived plastics and other non-renewable sources.
— = Not applicable.
Note: Totals may not equal sum of components due to independent rounding. Data for 2009 and 2010 are model results and may differ slightly from official EIA data reports.

SOURCE: Adapted from "Table A.16. Renewable Energy Generating Capacity and Generation," in *Annual Energy Outlook 2012*, U.S. Energy Information Administration, June 25, 2012, http://www.eia.gov/forecasts/aeo/pdf/0383(2012).pdf (accessed October 3, 2012)

World Production and Consumption

Hydropower is the most widely used renewable energy source in the world. In 2008 hydroelectric power accounted for 3,121 billion kWh, or 85% of total renewable electricity generation. (See Table 6.2.) The EIA predicts that hydroelectric power will increase to 5,620 billion kWh by 2035 but that its overall portion of total renewable electricity generation will decrease to 68%. The agency provides energy forecasts for countries that are and are not members of the Organisation for Economic Co-operation and Development (OECD). As explained in Chapter 5, the OECD is a collection of dozens of mostly western nations that are devoted to global economic development. OECD members are expected to make small gains in hydroelectric generation from 1,329 billion kWh in 2008 to 1,717 billion kWh in 2035, for a 1% annual growth rate. (See Table 6.2.) Nonmembers are projected to make larger generation additions. Their hydroelectric generation is projected to grow from 1,791 billion kWh in 2008 to 3,903 billion kWh in 2035, for a 2.9% annual growth rate.

FIGURE 6.6

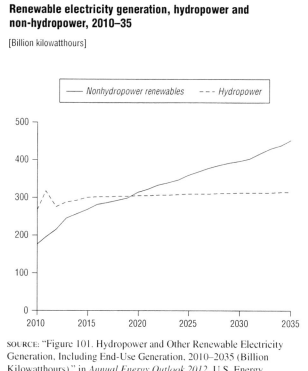

Renewable electricity generation, hydropower and non-hydropower, 2010–35

[Billion kilowatthours]

SOURCE: "Figure 101. Hydropower and Other Renewable Electricity Generation, Including End-Use Generation, 2010–2035 (Billion Kilowatthours)," in *Annual Energy Outlook 2012*, U.S. Energy Information Administration, June 25, 2012, http://www.eia.gov/forecasts/aeo/pdf/0383(2012).pdf (accessed October 3, 2012)

Most of the new development in hydropower is occurring in developing nations because they see it as an effective means of supplying power to growing populations. These massive public-works projects usually require huge amounts of money—most of it borrowed from the developed world. However, hydroelectric dams are considered worth the cost and potential environmental threats because they bring cheap electric power to the citizenry.

The Chinese government has constructed the world's largest dam, the Three Gorges Dam, on the Yangtze River in Hubei Province. Five times the size of the Hoover Dam in the United States, the Three Gorges Dam is 607 feet (185 m) tall and 7,575 feet (2,309 m) in length. One decade after the project was launched in 1993, the dam began generating power. The article "China's Three Gorges Dam Reaches Operating Peak" (BBC News, July 5, 2012) indicates that the dam reached its full operating peak in July 2012, after its 32nd, and final, generator began operating. The addition boosted the dam's generating capacity to 22.5 gigawatts, or around 11% of China's total hydropower capacity at that time. The article notes that the dam cost approximately $40 billion to construct and displaced more than 1 million people from the Yangtze River basin.

Environmental Issues

Dams have been used for decades in the United States for power generation, flood control, irrigation,

TABLE 6.2

World net renewable electricity generation, by source, 2008 and 2035

[Billion kilowatthours]

Region	2008	2035	Average annual percent change, 2008–2035
OECD			
Hydroelectric	1,329	1,717	1.0
Wind	181	898	6.1
Geothermal	38	104	3.8
Solar	12	120	8.8
Other	217	398	2.3
Total OECD	**1,778**	**3,236**	**2.2**
Non-OECD			
Hydroelectric	1,791	3,903	2.9
Wind	29	564	11.6
Geothermal	22	81	5.0
Solar	0	71	22.8
Other	41	375	8.5
Total non-OECD	**1,884**	**4,995**	**3.7**
World			
Hydroelectric	3,121	5,620	2.2
Wind	210	1,462	7.5
Geothermal	60	186	4.2
Solar	13	191	10.6
Other	258	772	4.1
Total world	**3,662**	**8,232**	**3.1**

Note: Totals may not equal sum of components due to independent rounding.
OECD = Organisation for Economic Co-opeation and Development.

SOURCE: Adapted from "Table 13. OECD and Non-OECD Net Renewable Electricity Generation by Energy Source, 2008–2035 (Billion Kilowatthours)," in *International Energy Outlook 2011*, U.S. Energy Information Administration, September 19, 2011, http://www.eia.gov/forecasts/ieo/pdf/0484(2011).pdf (accessed October 3, 2012).

and navigation. Hydroelectric power dams, in particular, have greatly disrupted natural water flows because they are so large. The structures also impede the migration paths of aquatic creatures, and their spinning turbines can kill and injure creatures unable to escape the blades. Even though modern dams typically include fish ladders (stepping-stone waterfalls that provide fish a pathway up and over dam structures), the dams still negatively affect aquatic life by altering natural water flows and temperatures.

Environmental concerns about dams are particularly acute in the Pacific Northwest because its rivers are home to many imperiled salmon species that migrate between freshwater and the sea. The Pacific Northwest and Northern California also contain many large hydroelectric generators. (See Figure 6.5.) The Klamath River has become the focal point of a debate over the environmental impacts of hydroelectric dams. It originates in southern Oregon and flows southwest into Northern California. The river is more than 200 miles (320 km) long and was once a major salmon habitat. However, the river has been heavily dammed, and those dams are believed to have degraded the habitat and harmed salmon

populations. Some of the dams were built decades ago, before fish ladders were required. Stephen Clark notes in "Study Shows Klamath Dam Removal Will Help Farmers, Fish but Skepticism Remains" (Fox News, September 24, 2011) that farmers and environmentalists have long fought in court over the best uses for the water from the Klamath River. A 2010 settlement reached between the sparring parties calls for the removal of four hydroelectric dams on the river, while a separate agreement specifies how river water will be allocated to competing demands, including maintaining thriving salmon populations. According to Clark, the dams are scheduled to be removed by 2020; however, it remains to be seen whether Congress will fund the project.

BIOMASS ENERGY

As noted earlier, *biomass* is a generic term that refers to various kinds of organic materials (other than fossil fuels) that can be used for energy production. Some analysts use the term *biomass* to refer only to solid organic materials, not liquids and gases.

The EIA divides biomass into three categories: wood, waste, and biofuels. The wood category includes wood and wood-derived fuels. The waste category includes municipal solid waste from biogenic (biologically derived) sources, landfill gas, sludge waste, agricultural by-products (such as stalks and husks), and other biomass sources. The biofuels category includes fuel ethanol (minus denaturant) and biodiesel. It also includes losses and co-products from the production of fuel ethanol and biodiesel. Fuel ethanol is ethanol intended for fuel use, rather than for human consumption. The EIA explains in *Annual Energy Review 2011* that fuel ethanol is denatured (made unfit for human consumption) by the addition of petroleum products. Fuel ethanol can be manufactured from various vegetative feedstocks, including starchy/sugary crops, such as corn, and cellulosic sources, such as trees and grasses. According to the DOE's Oak Ridge National Laboratory (ORNL), in *Biomass Energy Data Book* (September 2011, http://cta.ornl.gov/bedb/pdf/BEDB4_Full_Doc.pdf), 96.2% of domestically produced fuel ethanol in 2011 was made from corn.

According to the EIA, fuel ethanol is used primarily for blending at low concentrations into motor gasoline. For example, gasohol is a gasoline blend that contains up to 10% fuel ethanol by volume and can be used just like gasoline in conventional vehicles. At high concentrations, ethanol is blended with gasoline to make fuels that are suitable only for specially designed vehicles called alternative-fuel vehicles (AFVs). For example, E85 is the short name for a fuel that contains 85% ethanol and 15% motor gasoline. Biodiesel is a fuel made from biological sources, such as soybeans or animal fats, and is used in vehicles that ordinarily run on petroleum-derived diesel fuel.

Domestic Production and Consumption

Wood accounted for 22% of overall renewables consumption in 2011, followed by biofuels at 21% and waste at 5%. (See Figure 6.1.) Combined, these three sources accounted for nearly half (48%) of renewables consumption in 2011. Historical consumption amounts are shown in Figure 6.2 and Table 6.3. Wood has historically been the most popular of the biomass sources; however, its dominance is being threatened by biofuels. In 2011 wood consumption totaled 1,987 trillion Btu, compared with 1,947 trillion Btu for biofuels. Waste consumption was much lower at 477 trillion Btu. Among biomass sources, wood had the largest consumption percentage (45%), followed by biofuels (44%) and waste (11%).

According to the EIA, in *Annual Energy Review 2011*, biomass consumption by end-use sector in 2011 was as follows:

- Industrial—1,311 trillion Btu of wood, 172 trillion Btu of waste, and 17 trillion Btu of fuel ethanol

- Transportation—1,042 trillion Btu of fuel ethanol and 112 trillion Btu of biodiesel

- Residential—430 trillion Btu of wood

- Commercial—71 trillion Btu of wood, 36 trillion Btu of waste, and 3 trillion Btu of fuel ethanol

The electric power sector is also a consumer of biomass. In 2011 waste accounted for 269 trillion Btu of electric power, while wood accounted for 175 trillion Btu. (See Figure 6.7.) According to Figure 6.8, wood consumption in 2011 for electricity generation was nearly equal for the industrial and electric power sectors. (Some industrial facilities produce their own electricity.) However, little waste was used by industry in 2011 to generate electricity. (See Figure 6.9.)

FOCUS ON BIOFUELS. In *Annual Energy Review 2011*, the EIA notes that annual domestic production equals annual domestic consumption for all renewables except biofuels. Biofuels are heavily processed, and some losses and co-products result during processing. In addition, small amounts of biofuels are imported and exported. According to the EIA, in 2011 fuel ethanol feedstocks (corn and other biomass inputs) totaled 1,922 trillion Btu. Losses and co-products totaled 770 trillion Btu. After denaturant addition, the heat content of the final product was 1,182 trillion Btu. The EIA indicates that exports totaled 28.5 million barrels in 2011, while 3.1 million barrels were imported. Overall, fuel ethanol production in 2011 was 13.9 billion gallons (52.8 billion L), whereas consumption was 12.9 billion gallons (48.7 billion L). Consumption of fuel ethanol grew slowly for decades and then increased dramatically during the first decade of the 21st century. (See Figure 6.10.)

TABLE 6.3

Biomass energy consumption, by source, 1949–2011

[Trillion Btu]

Year	Biomass Wood[a]	Waste[b]	Biofuels[c]	Total
1949	1,549	NA	NA	1,549
1950	1,562	NA	NA	1,562
1955	1,424	NA	NA	1,424
1960	1,320	NA	NA	1,320
1965	1,335	NA	NA	1,335
1970	1,429	2	NA	1,431
1975	1,497	2	NA	1,499
1976	1,711	2	NA	1,713
1977	1,837	2	NA	1,838
1978	2,036	1	NA	2,038
1979	2,150	2	NA	2,152
1980	2,474	2	NA	2,476
1981	2,496	88	13	2,596
1982	2,510	119	34	2,663
1983	2,684	157	63	2,904
1984	2,686	208	77	2,971
1985	2,687	236	93	3,016
1986	2,562	263	107	2,932
1987	2,463	289	123	2,875
1988	2,577	315	124	3,016
1989	2,680	354	125	3,159
1990	2,216	408	111	2,735
1991	2,214	440	128	2,782
1992	2,313	473	145	2,932
1993	2,260	479	169	2,908
1994	2,324	515	188	3,028
1995	2,370	531	200	3,101
1996	2,437	577	143	3,157
1997	2,371	551	184	3,105
1998	2,184	542	201	2,927
1999	2,214	540	209	2,963
2000	2,262	511	236	3,008
2001	2,006	364	253	2,622
2002	1,995	402	303	2,701
2003	2,002	401	404	2,807
2004	2,121	389	499	3,010
2005	R2,137	403	577	R3,117
2006	R2,099	397	771	R3,267
2005	R2,070	413	991	R3,474
2008	R2,040	436	1,372	R3,849
2009	R1,891	R453	R1,568	R3,912
2010	R1,988	R469	R1,837	R4,294
2011	1,987	477	1,947	4,411

[a]Wood and wood-derived fuels.
[b]Municipal solid waste from biogenic sources, landfill gas, sludge waste, agricultural byproducts, and other biomass. Through 2000, also includes non-renewable waste (municipal solid waste from non-biogenic sources, and tire-derived fuels).
[c]Fuel ethanol (minus denaturant) and biodiesel consumption, plus losses and co-products from the production of fuel ethanol and biodiesel.
R = Revised. NA = Not available.
Btu = British thermal unit.
Note: Most data for the residential, commercial, industrial, and transportation sectors are estimates. Totals may not equal sum of components due to independent rounding.

SOURCE: Adapted from "Table 10.1. Renewable Energy Production and Consumption by Primary Energy Source, Selected Years, 1949–2011 (Trillion Btu)," in *Annual Energy Review 2011*, U.S. Energy Information Administration, September 27, 2012, http://www.eia.gov/totalenergy/data/annual/pdf/aer.pdf (accessed October 3, 2012)

According to the EIA, biodiesel production totaled 967 million gallons (3.7 billion L) in 2011, whereas consumption was 878 million gallons (3.3 billion L). Exports and imports were 1.7 million barrels and 861,000 barrels, respectively. The consumption of

FIGURE 6.7

Non-hydroelectric electric power consumption,1989–2011

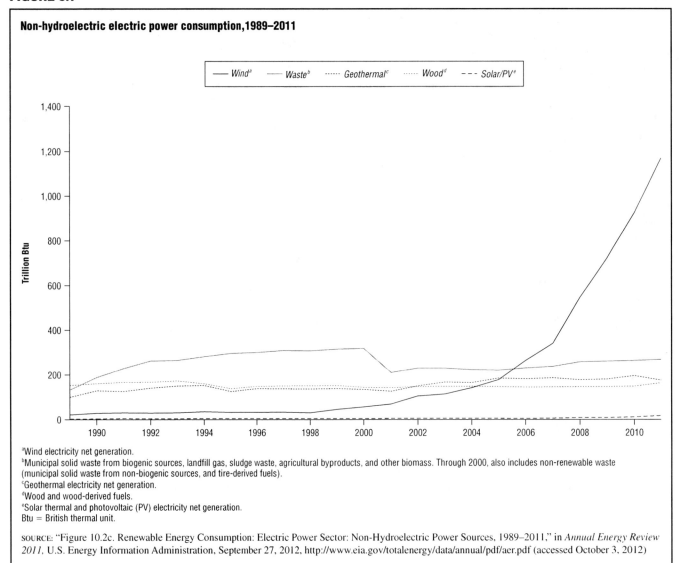

ᵃWind electricity net generation.
ᵇMunicipal solid waste from biogenic sources, landfill gas, sludge waste, agricultural byproducts, and other biomass. Through 2000, also includes non-renewable waste (municipal solid waste from non-biogenic sources, and tire-derived fuels).
ᶜGeothermal electricity net generation.
ᵈWood and wood-derived fuels.
ᵉSolar thermal and photovoltaic (PV) electricity net generation.
Btu = British thermal unit.

SOURCE: "Figure 10.2c. Renewable Energy Consumption: Electric Power Sector: Non-Hydroelectric Power Sources, 1989–2011," in *Annual Energy Review 2011*, U.S. Energy Information Administration, September 27, 2012, http://www.eia.gov/totalenergy/data/annual/pdf/aer.pdf (accessed October 3, 2012)

biodiesel was only around 300 million gallons (1.1 billion L) from 2007 to 2010 and then nearly tripled in 2011.

Domestic Market Factors

Wood and waste consumption were relatively flat between 1990 and 2011, whereas biofuels consumption skyrocketed. (See Table 6.3.) This market growth was driven largely by government mandates for biofuels use. Janet McGurty and Matthew Robinson report in "Analysis: U.S. Government Mandate or No, Fuel Ethanol Is Here to Stay" (Reuters, August 24, 2012) that in 1990 the Clean Air Act was amended to require the motor gasoline industry to sell reformulated gasoline (RFG) that burns cleaner (i.e., produces less air pollutants) than regular gasoline. RFG burns cleaner because it contains an oxygenate additive. Even though the RFG requirement originally applied to only a handful of U.S. cites with air pollution problems, by 2012 approximately 30% of the nation's gasoline stations were required to sell RFG. Ethanol has become the primary oxygenate used to produce RFG.

As noted earlier, the Energy Independence and Security Act of 2007 includes a Renewable Fuel Standard (RFS) that mandates biofuel use. Specifically, the RFS requires that by 2022 U.S. refineries will be blending 36 billion gallons (136.3 billion L) of biofuels into their production of transportation fuel annually. According to the DOE, in "Renewable Fuel Standard" (November 8, 2012, http://www.afdc.energy.gov/laws/RFS), the required RFS volumes will increase yearly, but the volumes may be changed by the U.S. Environmental Protection Agency based on actual biofuel availability. By 2022 the total is to consist of 16 billion gallons (60.6 billion L) of cellulosic ethanol, 15 billion gallons (56.8 billion L) of conventional (i.e., corn-derived) ethanol, 4 billion gallons (15.1 billion L) of advanced biofuels (e.g., sugar-derived ethanol), and 1 billion gallons (3.8 billion L) of biodiesel. According to McGurty and Robinson, the RFS required approximately 13.2 billion gallons (50 billion L) of ethanol consumption for blending in 2012, a goal that was expected to be met by the end of the year.

FIGURE 6.8

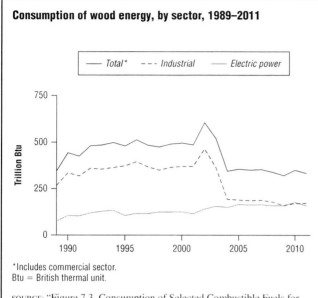

Consumption of wood energy, by sector, 1989–2011

*Includes commercial sector.
Btu = British thermal unit.

SOURCE: "Figure 7.3. Consumption of Selected Combustible Fuels for Electricity Generation: Wood by Sector, 1989–2011," in *Monthly Energy Review: September 2012*, U.S. Energy Information Administration, September 26, 2012, http://www.eia.gov/totalenergy/data/monthly/pdf/mer.pdf (accessed October 3, 2012)

FIGURE 6.9

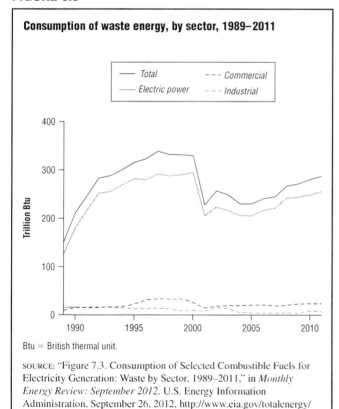

Consumption of waste energy, by sector, 1989–2011

Btu = British thermal unit.

SOURCE: "Figure 7.3. Consumption of Selected Combustible Fuels for Electricity Generation: Waste by Sector, 1989–2011," in *Monthly Energy Review: September 2012*, U.S. Energy Information Administration, September 26, 2012, http://www.eia.gov/totalenergy/data/monthly/pdf/mer.pdf (accessed October 3, 2012)

The biomass and biofuels industries also benefit from financial incentives that are provided by the federal government. The EIA indicates in *Direct Federal Financial*

FIGURE 6.10

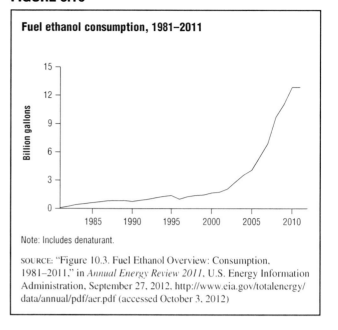

Fuel ethanol consumption, 1981–2011

Note: Includes denaturant.

SOURCE: "Figure 10.3. Fuel Ethanol Overview: Consumption, 1981–2011," in *Annual Energy Review 2011*, U.S. Energy Information Administration, September 27, 2012, http://www.eia.gov/totalenergy/data/annual/pdf/aer.pdf (accessed October 3, 2012)

Interventions and Subsidies in Energy in Fiscal Year 2010 that the biomass/biofuels industries received $7.8 billion in direct expenditures, tax breaks, R&D funds, and other means of financial support in 2010. This was more than half (53%) of the nearly $14.7 billion that was allocated for the renewables industry as a whole that year. Another $302 million went to "other" renewables, which the EIA explains included landfill gas, municipal solid waste, and hydrogen. Thus, the biomass industry received some of this funding as well.

Government incentives for fuel ethanol production have long been controversial because corn is a food crop. Critics fear that strong demand for corn-derived fuel ethanol since 1990 has driven up prices for food corn for humans, pets, and livestock. At the end of 2011 the federal government let a long-standing subsidy for corn ethanol expire. The article "Congress Ends Era of Ethanol Subsidies" (National Public Radio News, January 3, 2012) notes that the subsidy had been in place for three decades and amounted to around $20 billion total.

The Domestic Outlook

Table 6.1 shows EIA estimates of biomass usage for electricity generation in 2010 and predicted for 2035. Generation from municipal waste by the electric power sector and all end-use sectors was 18.6 billion kWh in 2010. It is expected to decrease to 17.5 billion kWh by 2035, for an annual rate of decline of 0.3%. Wood and other biomass was used to generate 37.6 billion kWh in 2010. Its use is projected to increase dramatically to 145.5 billion kWh in 2035, a growth rate of 5.6% annually.

According to the EIA, in *Annual Energy Outlook 2012*, 13.2 billion gallons (50 billion L) of ethanol were blended into gasoline in 2010, up from around 1.5 billion gallons (5.7 billion L) in 2000. The agency forecasts that ethanol consumption for blending will grow only slightly between 2010 and 2030, reaching 15.8 billion gallons (59.8 billion L) in 2035. Consumption of E85 fuel is expected to begin increasing by the latter half of the second decade of the 21st century as more AFVs come onto the market. By 2035 approximately 9.5 billion gallons (36 billion L) of ethanol are projected to be consumed annually for use in E85. However, the EIA predicts that the United States will not reach its RFS goal of 36 billion gallons (136.3 billion L) of biofuel consumption in 2022. The agency indicates that even though the corn-based ethanol and biodiesel mandates will likely be met, the mandates for cellulosic and advanced ethanol fuels will probably not be met because of "financial and technological hurdles."

The EIA also notes that in January 2011 the Environmental Protection Agency increased the maximum blending limit for ethanol in gasoline from 10% to 15% for vehicles built in 2001 and later. The EIA does not believe that the E15 mandate, which had not been finalized as of mid-2012, will be a major factor in future ethanol consumption. Furthermore, the agency is skeptical that refiners, retailers, and consumers will accept a higher ethanol blend for gasoline because of fears about potential damage to vehicle engines.

World Biomass Production

Biomass production figures on a worldwide basis are difficult to determine because country level data are not available for many countries. In addition, biomass makes up such a small share of energy production sources that in reports it is often lumped together with other renewables in an "other" category. However, some data for particular biomass components are available.

The ORNL indicates in *Biomass Energy Data Book* that world fuel ethanol production in 2010 was 22.9 billion gallons (86.7 billion L). The United States was, by far, the leading producer, with 13.2 billion gallons (50 billion L). Other major producers included Brazil (6.6 billion gallons [24.9 billion L]), the European Union (1 billion gallons [3.9 billion L]), China (541.6 million gallons [2 billion L]), and Canada (290.6 million gallons [1.1 billion L]).

According to the ORNL, world biodiesel production in 2009 was 308,200 barrels per day (bpd). Germany was the largest producer, accounting for 51,200 bpd. It was followed by France (41,100 bpd), the United States (32,900 bpd), Brazil (27,700 bpd), and Italy (13,100 bpd).

Environmental Issues

In general, biomass sources are hailed as more environmentally friendly than fossil fuels. However, the use of biomass is not without environmental problems. Deforestation can occur from widespread use of wood,

especially if forests are clear-cut, which can result in soil erosion and mudslides. In addition, burning biomass produces air emissions that degrade air quality. Although biomass is less energy intensive (i.e., has a lower heat content) than fossil fuels, combustion of any carbon-containing fuel releases carbon into the atmosphere that contributes to problems with global warming and climate change.

WIND ENERGY

Winds are created by the uneven heating of the atmosphere by the sun, the irregularities of the earth's surface, and the rotation of the planet. They are strongly influenced by bodies of water, weather patterns, vegetation, and other factors. The natural power of wind energy is collected by wind turbines, which resemble airplane propellers. The turbines convert wind energy to mechanical energy and finally to electrical energy. Figure 6.11 shows a wind turbine in which the blades rotate around a horizontal axis. Other types feature vertically aligned blades. Wind turbines are usually clustered on wind farms that may feature dozens or hundreds of individual wind turbines. The most favorable locations for wind turbines are in mountain passes and offshore along coastlines, where wind speeds are generally highest and most consistent.

FIGURE 6.11

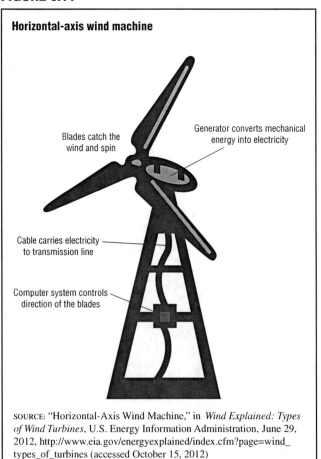

Horizontal-axis wind machine

Blades catch the wind and spin

Generator converts mechanical energy into electricity

Cable carries electricity to transmission line

Computer system controls direction of the blades

SOURCE: "Horizontal-Axis Wind Machine," in *Wind Explained: Types of Wind Turbines*, U.S. Energy Information Administration, June 29, 2012, http://www.eia.gov/energyexplained/index.cfm?page=wind_types_of_turbines (accessed October 15, 2012)

Domestic Production and Consumption

Wind energy accounted for 13% of total domestic renewables energy consumption in 2011. (See Figure 6.1.) Figure 6.2 shows historical wind energy consumption between 1949 and 2011. Wind was a small energy provider through the 1990s. At the turn of the 21st century consumption began to grow and increased dramatically through the latter part of the first decade, reaching 1.2 quadrillion Btu in 2011. Wind was the leading energy source for nonhydropower electricity generation in 2011. (See Figure 6.7.)

Domestic Market Factors

The wind industry, like other renewables industries, receives financial incentives from the federal government. In *Direct Federal Financial Interventions and Subsidies in Energy in Fiscal Year 2010*, the EIA notes that the wind industry received $5 billion in direct expenditures, tax breaks, R&D funds, loan guarantees, and other means of financial support in 2010. This was 34% of the $14.7 billion that was allocated for the renewables industry as a whole that year. Another $971 million in 2010 went to support electricity transmission and distribution. Thus, this funding also benefitted the wind industry. In addition, the state-level renewable portfolio standards described earlier help drive wind energy production for electricity generation.

Domestic Outlook

The EIA anticipates that wind energy will continue to dominate over solar, biomass, geothermal, municipal solid waste, and landfill gas for electricity generation for decades to come. (See Figure 6.12.) In 2010 wind provided 39 gigawatts of electricity. By 2035 it is forecast to provide nearly 70 gigawatts. Total electricity generation by wind for the electric power sector and all end-use sectors is projected to increase from 95 billion kWh in 2010 to 194.2 billion kWh in 2035, a growth rate of 2.9% annually. (See Table 6.1.)

World Production and Outlook

World wind energy production in 2008 was 210 billion kWh. (See Table 6.2.) By 2035 the EIA anticipates that wind will generate 1,462 billion kWh, an increase of 7.5% annually. OECD members are expected to make modest gains in wind power generation, from 181 billion kWh in 2008 to 898 billion kWh in 2035, for a 6.1% annual growth rate. Wind-power generation by nonmembers is projected to grow from 29 billion kWh in 2008 to 564 billion kWh in 2035, for an 11.6% annual growth rate.

Environmental Issues

In general, wind energy is considered to be environmentally friendly because it does not involve fuel combustion. Some people find the whirring noise of wind

FIGURE 6.12

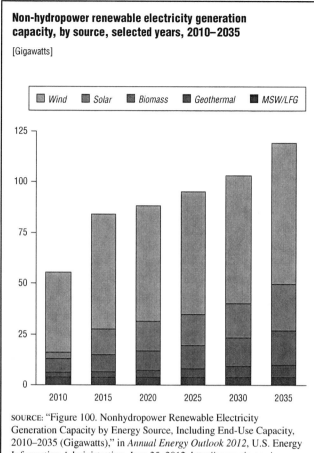

Non-hydropower renewable electricity generation capacity, by source, selected years, 2010–2035

[Gigawatts]

SOURCE: "Figure 100. Nonhydropower Renewable Electricity Generation Capacity by Energy Source, Including End-Use Capacity, 2010–2035 (Gigawatts)," in *Annual Energy Outlook 2012*, U.S. Energy Information Administration, June 25, 2012, http://www.eia.gov/forecasts/aeo/pdf/0383(2012).pdf (accessed October 3, 2012)

turbines annoying and object to clusters of wind turbines in mountain passes and along shorelines, where they interfere with scenic views. Environmentalists also point out that wind turbines are responsible for the loss of thousands of birds and bats that inadvertently fly into the blades. Birds frequently use windy passages in their travel patterns. However, wind farms do not emit climate-altering carbon dioxide and other pollutants, respiratory irritants, or radioactive waste. Furthermore, because wind farms do not require water to operate, they are especially well suited to semiarid and arid regions.

GEOTHERMAL ENERGY

Geothermal energy is the natural, internal heat of the earth trapped in rock formations deep underground. Only a fraction of it can be extracted, usually through large fractures in the earth's crust. Hot springs, geysers, and fumaroles (holes in or near volcanoes from which vapor escapes) are the most easily exploitable sources. Geothermal reservoirs provide hot water or steam that can be used for heating buildings and processing food. Pressurized hot water or steam can also be directed toward turbines, which spin, generating electricity for residential and commercial customers.

Geothermal energy accounted for 2% of total domestic renewables energy consumption in 2011. (See Figure 6.1.) Geothermal consumption was 163 trillion Btu that year. (See Figure 6.7.) The vast majority of the consumption was by the electric power sector. In *Annual Energy Review 2011*, the EIA notes that geothermal sources generated 16.7 billion kWh of net electricity generation in 2011, up slightly from 14.6 billion kWh in 1989. Thus, geothermal energy has been a very slow growing source of renewable energy. Greater use is limited by geological constraints. Geothermal energy is usable only when it is concentrated in one spot—in this case, in what is known as a thermal reservoir. Most of the known reservoirs for geothermal power in the United States are located west of the Mississippi River, and the highest-temperature geothermal resources occur for the most part west of the Rocky Mountains.

Domestic Market Factors

In *Direct Federal Financial Interventions and Subsidies in Energy in Fiscal Year 2010*, the EIA notes that the geothermal industry received $273 million in direct expenditures, tax breaks, R&D funds, loan guarantees, and other means of financial support in 2010. This was less than 2% of the $14.7 billion that was allocated for the renewables industry as a whole that year. As noted earlier, another $971 million in 2010 went to support electricity transmission and distribution. This funding may have also benefitted the geothermal industry. In addition, the state-level renewable portfolio standards may provide a boost to geothermal production and consumption for electricity generation.

The Domestic Outlook

Most of the easily exploited geothermal reserves in the United States have already been developed. Continued growth in the U.S. market depends on the regulatory environment, price trends for competing energy sources, and the success of new technologies to exploit previously inaccessible reserves. The EIA projects that geothermal energy generation will increase from 15.7 billion kWh in 2010 to 46.5 billion kWh in 2035, for an annual growth rate of 4.5%. (See Table 6.1.) All of the growth is forecast to occur in the electric power sector.

World Production and Consumption

Table 6.2 provides EIA estimates of worldwide geothermal-based electricity generation for 2008 and forecast for 2035. The agency notes that 60 billion kWh were generated in 2008. By 2035 the value is expected to reach 186 billion kWh, for a growth rate of 4.2%. OECD members are expected to boost their geothermal electricity generation from 38 billion kWh in 2008 to 104 billion kWh in 2035, for a 3.8% annual growth rate. Generation by nonmembers is projected to increase from 22 billion kWh in 2008 to 81 billion kWh in 2035, for a 5% annual growth rate.

World geothermal reserves are unevenly distributed. They occur mostly in seismically active areas at the margins or borders of the planet's nine tectonic plates. Areas that are rich in geothermal reserves include the west coasts of North and South America, Japan, the Philippines, and Indonesia.

SOLAR ENERGY

Solar energy, which comes from the sun, is a renewable, widely available energy source that does not generate pollution or radioactive waste. However, converting solar energy to electricity on a commercial scale has proved to be technologically and economically challenging.

There are two main types of solar systems: passive and active. In both systems the conversion of solar energy into a form of power is made at the site where it is used.

Passive solar energy systems, such as greenhouses or windows with a southern exposure, use heat flow, evaporation, or other natural processes to collect and transfer heat. They are considered to be the least costly and least difficult solar systems to implement.

Active solar systems require collectors and storage devices as well as motors, pumps, and valves to operate the systems that transfer heat. Some collectors consist of an absorbing plate that transfers the sun's heat to a working fluid (liquid or gas). Photovoltaic cells (which are combined to form solar rooftop panels or collectors) convert sunlight directly to electricity without the use of mechanical generators. Photovoltaic cells do not have moving parts, are easy to install, and require little maintenance. The use of photovoltaic cells is expanding around the world. Because they contain no turbines or other moving parts, operating costs are low and maintenance is minimal. Above all, the fuel source (sunshine) is free and plentiful. The main disadvantage of photovoltaic cell systems is the high initial cost, although prices have fallen considerably. Even though toxic materials are often used in the construction of the cells, researchers are investigating new materials, recycling, and disposal.

Domestic Production and Consumption

Solar/photovoltaic energy accounted for 2% of total domestics renewables consumption in 2011 and provided 158 trillion Btu. (See Figure 6.1.) This value has almost tripled since 1989, when it was 55 trillion Btu. Historically, most solar/photovoltaic energy has been consumed by the residential sector for electricity generation. According to the EIA, in *Annual Energy Review 2011*, solar/photovoltaic energy provided 18 trillion Btu to the electric power sector in 2011. This consumption was small in comparison to that of competing power sources. (See Figure 6.7.)

Domestic Market Factors

The EIA indicates in *Direct Federal Financial Interventions and Subsidies in Energy in Fiscal Year 2010* that the solar industry received $1.1 billion in direct expenditures, tax breaks, R&D funds, loan guarantees, and other means of financial support in 2010. This was around 8% of the $14.7 billion that was allocated for the renewables industry as a whole that year. In addition, the industry likely benefitted from another $971 million that went to support electricity transmission and distribution in 2010. Likewise, the state-level renewable portfolio standards enhance the production and consumption of solar power for electricity generation.

Domestic Outlook

Solar power's primary disadvantage is its reliance on a consistently sunny climate, which is possible in limited geographical areas. It also requires a large amount of land for the most efficient collection of solar energy by electricity plants.

Table 6.1 shows EIA estimates of solar energy electricity generation for 2010 and 2035. The agency believes that solar-generated power will increase from 4.5 billion kWh in 2010 to 44 billion kWh in 2035, for an annual growth rate of 9.6%. Strong growth is predicted for both the electric power sector and end-use sectors. Solar power is expected to grab an ever-increasing share of the renewables electricity generation market through 2035. (See Figure 6.12.)

World Production and Consumption

Table 6.2 provides EIA estimates of worldwide electricity generation from renewables for 2008 and predicted for 2035. The agency indicates that 13 billion kWh were generated from solar power in 2008. By 2035 the value is expected to reach 191 billion kWh, for a growth rate of 10.6%. OECD members are projected to increase their solar electricity generation from 12 billion kWh in 2008 to 120 billion kWh in 2035, for an 8.8% annual growth rate.

Generation by nonmembers was virtually zero in 2008. It is expected to rise to 71 billion kWh in 2035, for a 22.8% annual growth rate.

RENEWABLE ENERGY SOURCES FOR THE FUTURE

As of December 2012, numerous other renewable energy sources were being investigated for their potential to provide thermal power and/or electricity generation. They included:

- Tidal power—tidal power plants use the movement of water as it ebbs and flows to generate power. A minimum tidal range of 9 feet to 15 feet (2.7 m to 4.6 m) is generally considered necessary for an economically feasible plant. (The tidal range is the difference in height between consecutive high and low tides.)

- Wave energy—ocean wave power plants capture energy directly from surface waves or from pressure changes below the surface to generate electricity.

- Ocean thermal energy conversion—this source uses the temperature difference between warm surface water and the cooler water in the ocean's depths to power a heat engine into producing electricity. Ocean thermal energy conversion systems can be installed on ships, barges, or offshore platforms with underwater cables that transmit electricity to shore.

- Hydrogen—hydrogen is the lightest and most abundant chemical element, and an interesting fuel from an environmental point of view. Its combustion produces only water vapor, and it is entirely carbon free. Three-quarters of the mass of the universe is hydrogen, so in theory the supply is ample. However, the combustible form of hydrogen is a gas and is not found in nature. It must be made from other energy sources, such as fossil fuels. Hydrogen can be split from water, but the processes are either quite costly, require a great deal of energy, or both.

CHAPTER 7
ENERGY RESERVES—OIL, GAS, COAL, AND URANIUM

Energy resources are not the same as energy reserves. The former encompasses all deposits that exist, whereas the latter has a much narrower definition. They are deposits about which some specific information is known or can be estimated. Reserves are quantified using specific criteria that are technologically and/or economically based. One problem is that there are many different terms used by U.S. government agencies to refer to reserves, such as "proved," "unproved," "technically recoverable," "undiscovered," "conventional," and "continuous." Thus, one must be careful to specify what type of reserves are being discussed. At the federal level, there are two primary agencies that estimate and publish energy reserve amounts:

- U.S. Geological Survey (USGS)—according to the USGS, in "About the Energy Program" (May 15, 2012, http://energy.usgs.gov/GeneralInfo/Aboutthe EnergyProgram.aspx), its Energy Resources Program "conducts research and assessments on the location, quantity, and quality of mineral and energy resources, including the economic and environmental effects of resource extraction and use." The USGS notes that it evaluates numerous energy sources, including oil, natural gas, coal, coalbed methane, gas hydrates (methane gas trapped within solid crystals), geothermal resources, uranium, oil shale, bitumen, and heavy oil (a dense type of crude oil).

- Energy Information Administration (EIA) within the U.S. Department of Energy—the EIA collects data about reserves from a variety of sources, including other federal and state agencies, industry, and academia.

OIL AND NATURAL GAS

As explained in earlier chapters, the United States is highly dependent on oil and natural gas as energy sources. Combined, these two fuels accounted for 62% of the United States' primary energy consumption in 2011. (See Figure 1.10 in Chapter 1.) Thus, sufficient domestic reserves of these fuels are vitally important to the nation's economic well-being.

PROVED RESERVES

In *U.S. Crude Oil, Natural Gas, and Natural Gas Liquids Proved Reserves, 2010* (August 2012, http://www.eia.gov/naturalgas/crudeoilreserves/pdf/uscrudeoil .pdf), the EIA defines proved reserves as "those volumes of oil and natural gas that geologic and engineering data demonstrate with reasonable certainty to be recoverable in future years from known reservoirs under existing economic and operating conditions." It is important to understand that this definition encompasses both technical and economic considerations. Proved reserves are not simply deposits that are technically viable to extract, they must also be economically viable to extract. Changing economic conditions affect proved reserves estimates in that rising market prices push proved reserves estimates upward, whereas falling market prices push estimates downward.

The EIA provides reserves estimates as of year-end 2010 for crude oil and lease condensate and for wet natural gas. As noted in Chapter 2, lease condensate is a liquid recovered from natural gas at the well (the extraction point) and is generally blended with crude oil for refining. Wet natural gas was broadly defined in Chapter 3, but the EIA uses a narrower definition in *U.S. Crude Oil, Natural Gas, and Natural Gas Liquids Proved Reserves, 2010*. It defines wet natural gas as including only natural gas plant liquids, which is a mixture of liquid compounds that are recovered during processing. The EIA indicates that it bases its proved reserves estimates in large part on the data it collects from annual surveys of approximately 1,200 domestic oil and gas operators.

According to the EIA, proved reserves of the fuels at year-end 2010 were:

- Crude oil and lease condensate—25.2 billion barrels

- Wet natural gas—317.6 trillion cubic feet (tcf [9 trillion cubic meters (tcm)])

Figure 7.1 shows historical data for proved reserves of crude oil and lease condensate dating back to 1980. The U.S. total fell from a high of 31 billion barrels in 1980 to a low of approximately 20 billion barrels in 2008. Subsequent estimates indicate greater reserve amounts. The uptick in total proved reserves in 2009 and 2010 reflects additional reserves that were estimated for onshore reservoirs in the lower 48 states. Figure 7.2 shows proved reserves volumes in 2010 by geographical area. Texas had the largest volume (6,356 million barrels), followed by the Gulf of Mexico federal offshore area (4,347 million barrels), California (2,939 million barrels), and North Dakota (1,887 million barrels).

Historical proved reserves data for wet natural gas are shown in Figure 7.3. The overall U.S. total was relatively flat or slightly declining from 1980 through the 1990s. A tremendous upswing then occurred. Proved reserves increased from around 175 tcf (5 tcm) in 1999 to nearly 320 tcf (9.1 tcm) in 2010. All of the increase was due to greater proved reserves in natural gas reservoirs in the lower 48 states. Figure 7.4 shows proved reserves of wet natural gas in 2010 by geographical area. Texas had, by far, the largest volume, at 94,287 billion cubic feet (bcf; 2,670 billion cubic meters [bcm]). Wyoming was the second largest, at 36,526 bcf [1,034 bcm], followed by Louisiana, at 29,517 bcf (835.8 bcm); Oklahoma, at 28,182 bcf (798 bcm); and Colorado, at 25,372 bcf (718.5 bcm).

The calculations involved in estimating proved reserves of both fuels between 2001 and 2010 are shown in Table 7.1. The EIA notes that "reserves estimates change from year to year as new discoveries are made, existing fields are more thoroughly appraised, existing reserves are produced, and as prices and technologies change."

The EIA indicates that recent reserves gains reflect higher estimates for unconventional geological sources (e.g., shale and tight formations). These sources are described in detail in Chapter 2 for oil and Chapter 3 for natural gas. Technological advancements, such as horizontal drilling and hydraulic fracturing, have allowed operators greater access to fuels that are trapped in formerly

FIGURE 7.1

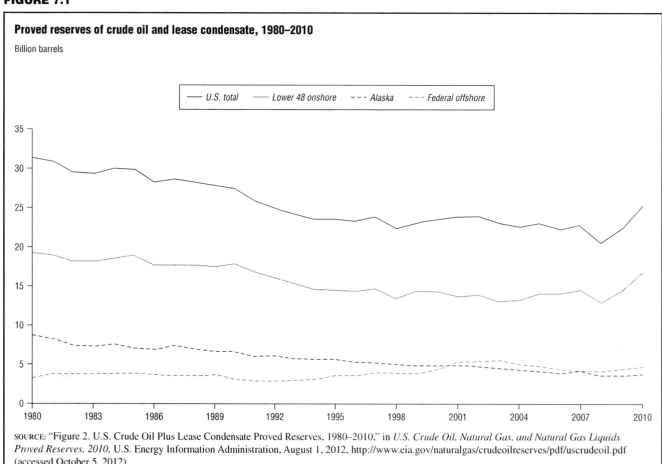

Proved reserves of crude oil and lease condensate, 1980–2010

Billion barrels

—— U.S. total   —— Lower 48 onshore   - - - Alaska   - - - Federal offshore

SOURCE: "Figure 2. U.S. Crude Oil Plus Lease Condensate Proved Reserves, 1980–2010," in *U.S. Crude Oil, Natural Gas, and Natural Gas Liquids Proved Reserves, 2010*, U.S. Energy Information Administration, August 1, 2012, http://www.eia.gov/naturalgas/crudeoilreserves/pdf/uscrudeoil.pdf (accessed October 5, 2012)

FIGURE 7.2

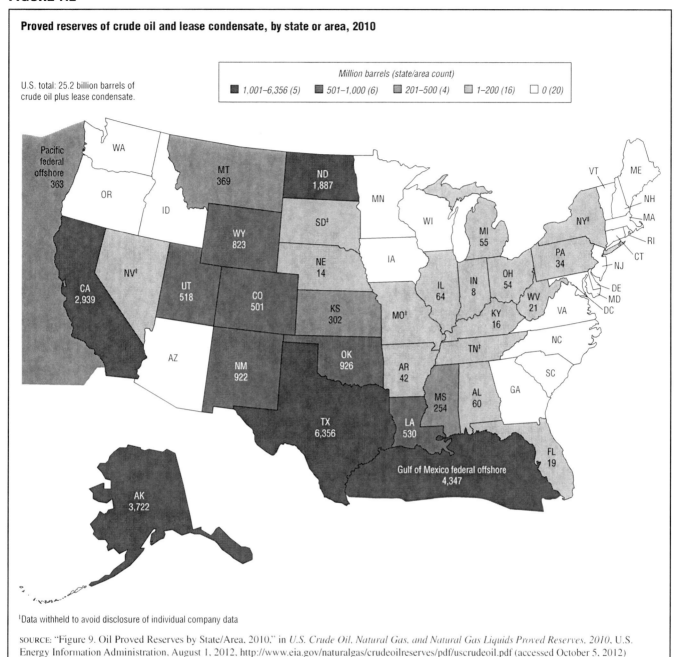

Proved reserves of crude oil and lease condensate, by state or area, 2010

Million barrels (state/area count)

■ 1,001–6,356 (5) ▨ 501–1,000 (6) ▦ 201–500 (4) ▢ 1–200 (16) ☐ 0 (20)

U.S. total: 25.2 billion barrels of crude oil plus lease condensate.

Pacific federal offshore 363

WA

MT 369

ND 1,887

MN

OR

ID

WY 823

SD‡

WI

MI 55

VT   ME

NY‡

NH

MA

RI

CT

NJ

PA 34

NV‡

UT 518

CO 501

NE 14

IA

IL 64

IN 8

OH 54

WV 21

VA

DE

MD

DC

CA 2,939

KS 302

MO‡

KY 16

NC

AZ

NM 922

OK 926

AR 42

TN‡

SC

TX 6,356

MS 254

AL 60

GA

LA 530

FL 19

AK 3,722

Gulf of Mexico federal offshore 4,347

‡Data withheld to avoid disclosure of individual company data

SOURCE: "Figure 9. Oil Proved Reserves by State/Area, 2010," in *U.S. Crude Oil, Natural Gas, and Natural Gas Liquids Proved Reserves, 2010*, U.S. Energy Information Administration, August 1, 2012, http://www.eia.gov/naturalgas/crudeoilreserves/pdf/uscrudeoil.pdf (accessed October 5, 2012)

inaccessible geological sources. In addition, the agency indicates that proved oil reserves increased in 2010 partly because of higher market prices for oil. As noted earlier, economic viability is a key consideration when estimating proved reserves volumes.

Some of the terms listed in Table 7.1 require more explanation. The column labeled "Total discoveries" is the sum of the columns labeled "Extensions," "New field discoveries," and "New reservoir discoveries in old fields." The EIA states that extensions are "reserve additions that result from additional drilling and exploration in previously discovered reservoirs." As shown in Table 7.1, extensions accounted for the vast majority of total discoveries between 2001 and 2010.

TECHNICALLY RECOVERABLE RESOURCES

Technically recoverable resources (TRR) are energy resources that are technically recoverable; however, it may or may not be economically viable to recover them. In "Geology and Technology Drive Estimates of Technically Recoverable Resources" (July 20, 2012, http://www.eia.gov/todayinenergy/detail.cfm?id=7190), the EIA notes that TRR is "a common measure of the long-term viability of U.S. domestic crude oil and natural gas as an energy source." However, the EIA warns in *Annual Energy Outlook 2012* (June 2012, http://www.eia.gov/forecasts/aeo/pdf/0383(2012).pdf) that TRR estimates are "highly uncertain" and change as new information is collected. The

FIGURE 7.3

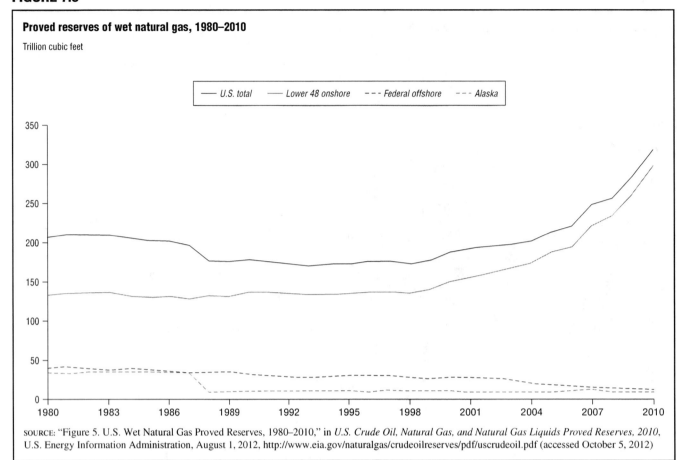

Proved reserves of wet natural gas, 1980–2010

Trillion cubic feet

Legend: —— U.S. total   —— Lower 48 onshore   - - - Federal offshore   - - - Alaska

SOURCE: "Figure 5. U.S. Wet Natural Gas Proved Reserves, 1980–2010," in *U.S. Crude Oil, Natural Gas, and Natural Gas Liquids Proved Reserves, 2010*, U.S. Energy Information Administration, August 1, 2012, http://www.eia.gov/naturalgas/crudeoilreserves/pdf/uscrudeoil.pdf (accessed October 5, 2012)

agency explains that it makes its estimates based on well production data and information gathered from other government agencies (both federal and state), industry, and academia. TRR includes both proved reserves (as defined earlier) and unproved resources, which the EIA defines as "additional volumes estimated to be technically recoverable without consideration of economics or operating conditions, based on the application of current technology."

In *Annual Energy Review 2011* (September 2012, http://www.eia.gov/totalenergy/data/annual/pdf/aer.pdf), the EIA indicates that in 2009 TRR for crude oil and lease condensate totaled 220.2 billion barrels broken down by geographical area as follows: onshore lower 48 states (126.7 billion barrels, or 58% of total), offshore lower 48 states (54.8 billion barrels, or 25% of total), and Alaska (38.6 billion barrels, or 18% of total). TRR for dry natural gas in 2009 totaled 2,203 tcf (62.4 tcm) broken down by type and geographical location as follows: unconventional natural gas, meaning tight gas, shale gas, and coalbed methane (1,193.8 tcf [33.8 tcm], or 54% of total), conventional natural gas onshore lower 48 states (451.1 tcf [12.8 tcm], or 20% of total), conventional natural gas offshore lower 48 states (277.6 tcf [7.9 tcm], or 13% of total), and conventional natural gas Alaska (280.8 tcf [8 tcm], or 13% of total). Overall,

unproved resources accounted for 90% of crude oil and lease condensate TRR and 88% of dry natural gas TRR.

It should be noted that the EIA does not include in its TRR estimates any oil and natural gas resources that are located in areas in which drilling is officially prohibited. In addition, the agency excludes resources in the northern Atlantic Outer Continental Shelf (OCS), the northern and central Pacific OCS, and within a 50-mile (80-km) buffer off the southern and mid-Atlantic OCS.

Off-Limit Areas

As of December 2012, there were several areas around the United States that were off-limits for oil and natural gas drilling for various reasons. Perhaps the most well-known (and most controversial) area is the Arctic National Wildlife Refuge (ANWR) in Alaska. The state's northern region, or North Slope, has long been a prolific production area for oil and natural gas. However, for decades the adjacent ANWR has been closed by federal law to oil and natural gas development. This 19-million-acre (7.7-million-ha) area of pristine wilderness lies along the Alaskan-Canadian border. The USGS believes that there are substantial volumes of oil and natural gas beneath ANWR. Oil from the North Slope is transported via the Trans-Alaska Pipeline System to the port city of

FIGURE 7.4

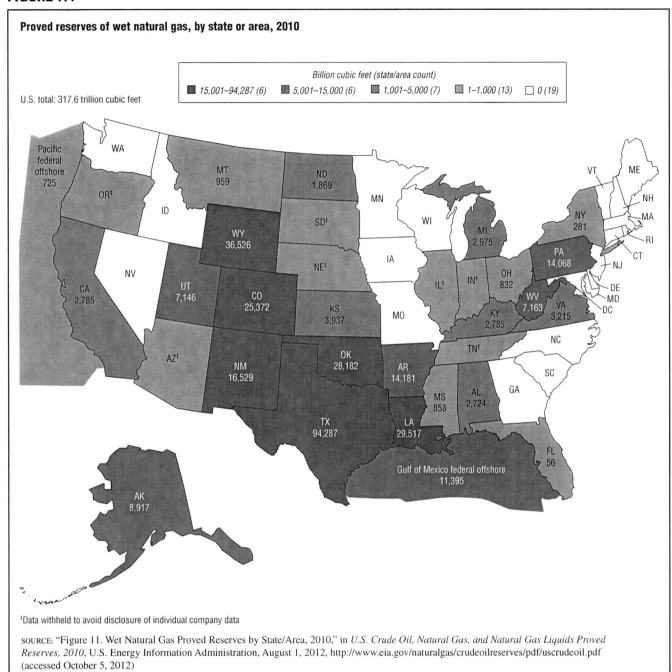

Proved reserves of wet natural gas, by state or area, 2010

U.S. total: 317.6 trillion cubic feet

Billion cubic feet (state/area count)

- 15,001–94,287 (6)
- 5,001–15,000 (6)
- 1,001–5,000 (7)
- 1–1,000 (13)
- 0 (19)

‡Data withheld to avoid disclosure of individual company data

SOURCE: "Figure 11. Wet Natural Gas Proved Reserves by State/Area, 2010," in *U.S. Crude Oil, Natural Gas, and Natural Gas Liquids Proved Reserves, 2010*, U.S. Energy Information Administration, August 1, 2012, http://www.eia.gov/naturalgas/crudeoilreserves/pdf/uscrudeoil.pdf (accessed October 5, 2012)

Valdez in southern Alaska. Over the years there have been numerous calls (mostly from Republican politicians) for ANWR to be opened to oil and gas drilling. However, as of December 2012, no federal legislation had been passed that would permit such development.

As explained in Chapter 1, the federal and state governments control resource leases for public lands within their jurisdictions, including offshore areas. Even though oil and natural gas drilling are prolific in the Gulf of Mexico, the same is not true for areas offshore the Atlantic and Pacific coasts. For decades, these areas have been off-limits to resource development. New offshore drilling along the West Coast was banned by the federal government and

the coastal states decades ago. In 2010 President Barack Obama (1961–) announced a controversial plan to open the Atlantic OCS to drilling. However, the plan was put on hold only months later after the British Petroleum oil spill occurred in the Gulf of Mexico. (See Chapter 2 for a description of the spill and its aftermath.) Thus, as of December 2012, no new oil and natural gas leases were in effect within U.S. waters off the Atlantic and Pacific coasts.

Exploration and Development

Finding oil and gas usually takes two steps. First, geological and geophysical exploration identifies areas where oil and gas are most likely to be found. Much of

TABLE 7.1

Proved reserves of wet natural gas and crude oil plus lease condensate, 2001–10

Year	Adjustments[a] (1)	Net revisions (2)	Revisions and adjustments[a] (3)	Net of sales[b] and acquisitions (4)	Extensions (5)	New field discoveries (6)	New reservoir discoveries in old fields (7)	Total[c] discoveries (8)	Estimated production (9)	Proved[d] reserves 12/31 (10)	Change from prior year (11)
Wet natural gas (billion cubic feet, 14.73 psia, 60 degrees Fahrenheit)											
2001	1,849	−2,438	−589	2,715	17,183	3,668	2,898	23,749	20,642	191,743	5,233
2002	4,006	1,038	5,044	428	15,468	1,374	1,752	18,594	20,248	195,561	3,818
2003	2,323	−1,715	608	1,107	17,195	1,252	1,653	20,100	20,231	197,145	1,584
2004	170	825	995	1,975	19,068	790	1,244	21,102	20,017	201,200	4,055
2005	1,693	2,715	4,408	2,674	22,069	973	1,243	24,285	19,259	213,308	12,108
2006	946	−2,099	−1,153	3,178	22,834	425	1,197	24,456	19,373	220,416	7,108
2007	990	15,936	16,926	452	28,255	814	1,244	30,313	20,318	247,789	27,373
2008	271	−3,254	−2,983	937	27,800	1,229	1,678	30,707	21,415	255,035	7,246
2009	5,923	−1,899	4,024	−222	43,500	1,423	2,656	47,579	22,537	283,879	28,844
2010	1,292	4,055	5,347	2,766	46,283	895	1,701	48,879	23,224	317,647	33,768
Crude oil plus lease condensate (million barrels of 42 U.S. gallons)											
2001	−61	−346	−407	−53	1,002	1,480	358	2,840	2,133	23,843	326
2002	423	682	1,105	51	600	318	187	1,105	2,082	24,023	180
2003	192	−9	183	−416	530	717	137	1,384	2,068	23,106	−917
2004	80	444	524	37	731	36	159	926	2,001	22,592	−514
2005	237	558	795	327	946	209	57	1,212	1,907	23,019	427
2006	109	43	152	189	685	38	62	785	1,834	22,311	−708
2007	21	1,275	1,296	44	865	81	87	1,033	1,872	22,812	501
2008	318	−2,189	−1,871	187	968	166	137	1,271	1,845	20,554	−2,258
2009	46	2,008	2,054	95	1,305	141	95	1,541	1,929	22,315	1,761
2010	188	1,943	2,131	667	1,766	124	169	2,059	1,991	25,181	2,866

[a]Revisions and adjustments = Col. 1 + Col. 2.
[b]Net of sales and acquisitions = acquisitions − sales
[c]Total discoveries = Col. 5 + Col. 6 + Col. 7.
[d]Proved reserves = Col. 10 from prior year + Col. 3 + Col. 4 + Col. 8 − Col. 9.
NA = Not available
Notes: Old means discovered in a prior year. New means discovered during the report year.

SOURCE: "Table 4. Total U.S. Proved Reserves of Wet Natural Gas, and Crude Oil Plus Lease Condensate, 2001–2010," in *U.S. Crude Oil, Natural Gas, and Crude Oil Plus Lease Condensate, 2001–2010;* in *U.S. Crude Oil, Natural Gas, and Natural Gas Liquids Proved Reserves, 2010,* U.S. Energy Information Administration, August 1, 2012. http://www.eia.gov/naturalgas/crudeoilreserves/pdf/uscrudeoil.pdf (accessed October 5, 2012)

this exploration is seismic, in which shock waves are used to determine the formations below the surface of the earth. Different rock formations transmit shock waves at different velocities, so they help determine if the geological features most often associated with oil and gas accumulations are present. After the seismic testing has been completed—and if it has been successful—exploratory wells are drilled. In *Annual Energy Review 2011*, the EIA notes that exploratory wells are drilled for three purposes: to find crude oil or natural gas in an area previously considered to be unproductive, to find a new reservoir in a field that has previously produced crude oil or natural gas in another reservoir, and to extend the limit of a crude oil or natural gas reservoir that has previously been productive. By contrast, the agency indicates that development wells are drilled within proved areas of reservoirs to depths that are known to be productive.

Table 7.2 provides historical data about exploratory and development well drilling in the United States. In 2010, 37,504 wells were drilled, including 16,254 crude oil wells and 16,973 natural gas wells. Another 4,277 wells were so-called dry holes, meaning that the wells could not produce either crude oil or natural gas "in sufficient quantities to justify completion" of the wells. Overall, 88.6% of the exploratory and development wells drilled in 2010 were successful (i.e., not dry holes). As shown in Table 7.2, fewer exploratory and development wells were drilled per year during the first decade of the 21st century than in some years in the past; however, the total footage drilled was generally higher. In 2010, 100.7 million feet (30.7 million m) of crude oil wells were drilled. For natural gas, the value was 147 million feet (44.8 million m). The average footage drilled in 2010 was 6,194 feet (1,888 m) per crude oil well and 8,659 feet (2,639 m) per natural gas well. In general, the average footage drilled per exploratory and development well has increased over the decades.

COAL

Coal supplied 20% of the nation's primary energy consumption in 2011. (See Figure 1.10 in Chapter 1.) The different ranks of coal (anthracite, bituminous coal, subbituminous coal, and lignite) are described in detail in Chapter 4. The EIA measures U.S. coal reserves in terms of the demonstrated reserve base (DRB). In *Annual Energy Review 2011*, the agency notes that the DRB sums the remaining "measured and indicated" coal resources in the United States that meet minimum seam and depth criteria.

Table 7.3 shows DRB data as of January 1, 2011. The total DRB was 484.5 billion tons (439.5 billion t), including 231 billion tons (209.6 billion t) located east of the Mississippi River and 253.5 billion tons (230 billion t) located west of the Mississippi River. Most (331.2 billion tons [300.5 billion t], or 68%) of the total reserves are expected to be extracted via underground mining methods, while the other 32% (153.3 billion tons [139.1 billion t]) are expected to be extracted using surface mining methods. Bituminous and subbituminous coal accounted for the largest portions of the total DRB. Lignite and anthracite reserves were much smaller.

On a regional and state basis the western region held the largest fraction of the DRB as of January 1, 2011, with Montana leading all the states, with 119 billion tons (108 billion t). It was followed by Illinois, with 104.2 billion tons (94.5 billion t), and Wyoming, with 61 billion tons (55.3 billion t).

The EIA indicates that it believes approximately 54% of the DRB is recoverable. However, the agency explains that recoverability using "current mining technologies" varies considerably by location.

URANIUM

As explained in Chapter 5, uranium is the resource used to produce nuclear power for electricity generation. The world's uranium supply is enormous because the element is present at low levels throughout the earth's crust and oceans. In fact, the Department of Energy's Oak Ridge National Laboratory reports in the press release "ORNL Technology Moves Scientists Closer to Extracting Uranium from Seawater" (August 21, 2012, http://www.ornl.gov/info/press_releases/get_press_release.cfm?ReleaseNumber=mr20120821-00) that its scientists have developed materials that may one day be used commercially to extract uranium from seawater. In the meantime, ores (bodies of minerals) are the primary resource for uranium extraction.

According to Figure 1.10 in Chapter 1, nuclear electric power accounted for 8% of the nation's primary energy consumption in 2011. In *Annual Energy Review 2011*, the EIA notes that it categorizes domestic uranium reserves using a common international category: reasonably assured resources (RAR). These reserves are known mineral deposits of uranium (i.e., uranium ores) of specific "size, grade, and configuration" and are recoverable "within the given production cost ranges, with currently proven mining and processing technology." Thus, uranium reserves are quantified using both technological and economic criteria.

The EIA indicates that in 2008 domestic uranium reserves (as uranium oxide) totaled 1.8 billion pounds (816 million kg). The largest reserves were in Wyoming (666 million pounds [302 million kg], or 38% of the total) and New Mexico (569 million pounds [258 million kg], or 32% of the total). Other states with reserves included Alaska, Arizona, California, Colorado, Idaho, Montana,

TABLE 7.2

Crude oil and natural gas exploratory and development wells drilled, selected years, 1949–2010

Year	Wells drilled (Number)				Successful wells (Percent)	Footage drilled (Thousand feet)				Average footage drilled (Feet per well)			
	Crude oil	Natural gas	Dry holes	Total		Crude oil	Natural gas	Dry holes	Total	Crude oil	Natural gas	Dry holes	Total
1949	21,352	3,363	12,597	37,312	66.2	79,428	12,437	43,754	135,619	3,720	3,698	3,473	3,635
1950	23,812	3,439	14,799	42,050	64.8	92,695	13,685	50,977	157,358	3,893	3,979	3,445	3,742
1955	30,432	4,266	20,452	55,150	62.9	121,148	19,930	85,103	226,182	3,981	4,672	4,161	4,101
1960	22,258	5,149	18,212	45,619	60.1	86,568	28,246	77,361	192,176	3,889	5,486	4,248	4,213
1965	18,065	4,482	16,226	38,773	58.2	73,322	24,931	76,629	174,882	4,059	5,562	4,723	4,510
1970	12,968	4,011	11,031	28,010	60.6	56,859	23,623	58,074	138,556	4,385	5,860	5,265	4,943
1975	16,948	8,127	13,646	38,721	64.8	66,819	44,454	69,220	180,494	3,943	5,470	5,073	4,661
1976	17,688	9,409	13,758	40,855	66.3	68,892	49,113	68,977	186,982	3,895	5,220	5,014	4,577
1977	18,745	12,122	14,985	45,852	67.3	75,451	63,686	76,728	215,866	4,025	5,254	5,120	4,708
1978	19,181	14,413	16,551	50,145	67.0	77,041	75,841	85,788	238,669	4,017	5,262	5,183	4,760
1979	20,851	15,254	16,099	52,204	69.2	82,688	80,468	81,642	244,798	3,966	5,275	5,071	4,689
1980	32,959	17,461	20,785	71,205	70.8	125,262	92,106	99,575	316,943	3,801	5,275	4,791	4,451
1981	43,887	20,250	27,953	92,090	69.6	172,167	108,353	134,934	415,454	3,923	5,351	4,827	4,511
1982	39,459	19,076	26,379	84,914	68.9	149,674	107,149	123,746	380,569	3,793	5,617	4,691	4,482
1983	37,366	14,684	24,355	76,405	68.1	136,849	78,108	105,222	320,179	3,662	5,319	4,320	4,191
1984	42,906	17,338	25,884	86,128	69.9	162,653	91,480	119,860	373,993	3,791	5,276	4,631	4,342
1985	35,261	14,324	21,211	70,796	70.0	137,728	76,293	100,388	314,409	3,906	5,326	4,733	4,441
1986	19,213	8,599	12,799	40,611	68.5	76,825	45,039	60,961	182,825	3,999	5,238	4,763	4,502
1987	16,210	8,096	11,167	35,473	68.5	66,358	42,584	53,588	162,530	4,094	5,260	4,799	4,582
1988	13,646	8,578	10,119	32,343	68.7	58,639	45,363	52,517	156,519	4,297	5,288	5,190	4,839
1989	10,230	9,522	8,236	27,988	70.6	43,266	49,081	42,099	134,446	4,229	5,154	5,112	4,804
1990	12,839	11,246	8,245	32,330	74.5	56,591	R57,028	42,433	R156,052	R4,408	R5,071	R5,147	R4,827
1991	12,588	9,793	7,481	29,862	74.9	56,196	R51,032	37,750	R144,978	R4,464	R5,211	R5,046	R4,855
1992	9,402	8,163	5,862	23,427	75.0	45,748	44,727	29,451	R119,926	R4,866	5,479	R5,024	R5,119
1993	8,856	9,839	6,096	24,791	75.4	44,236	R58,240	31,018	R133,494	R4,995	R5,919	R5,088	R5,385
1994	7,348	9,375	5,096	21,819	76.6	38,620	58,340	27,771	R124,731	R5,256	6,223	R5,450	R5,717
1995	8,248	8,082	4,814	21,144	77.2	41,076	49,746	26,349	R117,171	R4,980	6,155	R5,473	R5,542
1996	8,836	9,027	4,890	22,753	78.5	42,472	56,042	27,851	R126,365	R4,807	6,208	R5,696	R5,554
1997	11,206	11,498	5,874	28,578	79.4	56,371	71,270	33,640	R161,281	R5,030	6,198	R5,727	R5,644
1998	7,682	11,639	4,761	24,082	80.2	38,579	70,099	28,540	R137,218	R5,022	6,023	R5,995	R5,698
1999	4,805	12,027	3,550	20,382	82.6	22,024	60,217	20,608	R102,849	4,584	5,007	R5,805	R5,046
2000	8,090	17,051	4,146	29,287	85.8	36,745	R83,618	24,076	R144,439	R4,542	R4,904	R5,807	R4,932
2001	8,888	22,072	4,598	35,558	87.1	43,172	R110,734	26,221	R180,127	R4,857	R5,017	R5,703	R5,066
2002	6,775	17,342	3,754	27,871	86.5	30,892	93,041	21,232	R145,165	R4,560	5,365	R5,656	R5,208
2003	8,129	20,722	3,982	32,833	87.9	38,588	R115,916	22,744	R177,248	R4,747	5,594	R5,712	R5,398
2004	8,789	24,186	4,082	37,057	89.0	42,109	R138,449	23,714	R204,272	R4,791	5,724	R5,809	R5,512
2005	10,779	R28,590	4,653	R44,022	89.4	51,449	R163,820	25,044	R240,313	R4,773	5,730	R5,382	R5,459
2006	R13,404	R32,838	R5,206	R51,448	89.9	R63,340	R191,646	R27,778	R282,764	R4,725	5,836	R5,336	R5,496
2007	R13,361	R32,719	R4,978	R51,058	90.3	R64,792	R208,907	R27,754	R301,453	R4,849	R6,385	R5,575	R5,904
2008	16,645	R32,347	R5,428	R54,347	90.0	82,646	R223,224	28,572	R334,442	R4,965	R6,917	R5,264	R6,154
2009	R11,261	R18,234	R3,552	33,047	89.3	R62,771	156,200	20,520	R239,491	R5,574	8,566	R5,777	R7,247
2010	R16,254	R16,973	R4,277	37,504	88.6	R100,682	146,973	21,719	R269,374	R6,194	8,659	R5,078	R7,183

R = Revised.

Notes: 2011 data for this table were not available in time for publication. Data are estimates. Data are for exploratory and development wells combined.
Service wells, stratigraphic tests, and core tests are excluded.
For 1949–1959, data represent wells completed in a given year. For 1960–1969, data are for well completion reports received by the American Petroleum Institute during the reporting year. For 1970 forward, the data represent wells completed in a given year. The as-received well completion data for recent years are incomplete due to delays in the reporting of wells drilled. The U.S. Energy Information Administration (EIA) therefore statistically imputes the missing data. Totals may not equal sum of components due to independent rounding. Average depth may not equal average of components due to independent rounding.

SOURCE: "Table 4.5. Crude Oil and Natural Gas Exploratory and Development Wells, Selected Years, 1949–2010," in *Annual Energy Review 2011,* U.S. Energy Information Administration, September 27, 2012, http:// www.eia.gov/totalenergy/data/annual/pdf/aer.pdf (accessed October 3, 2012)

TABLE 7.3

Coal demonstrated reserve base, as of January 1, 2011

[Billion short tons]

Region and state	Anthracite Underground	Anthracite Surface	Bituminous coal Underground	Bituminous coal Surface	Subbituminous coal Underground	Subbituminous coal Surface	Lignite Surface[a]	Total Underground	Total Surface	Total
Appalachian	**4.0**	**3.3**	**68.2**	**21.9**	**0.0**	**0.0**	**1.1**	**72.1**	**26.3**	**98.4**
Alabama	0.0	0.0	0.9	2.1	0.0	0.0	1.1	0.9	3.1	4.0
Kentucky, Eastern	0.0	0.0	0.8	9.1	0.0	0.0	0.0	0.8	9.1	9.8
Ohio	0.0	0.0	17.4	5.7	0.0	0.0	0.0	17.4	5.7	23.1
Pennsylvania	3.8	3.3	18.9	0.8	0.0	0.0	0.0	22.7	4.2	26.9
Virginia	0.1	0.0	0.9	0.5	0.0	0.0	0.0	1.0	0.5	1.5
West Virginia	0.0	0.0	28.3	3.4	0.0	0.0	0.0	28.3	3.4	31.7
Other[b]	0.0	0.0	1.1	0.3	0.0	0.0	0.0	1.1	0.3	1.4
Interior	**0.1**	**(s)**	**116.6**	**27.1**	**0.0**	**0.0**	**12.6**	**116.7**	**39.6**	**156.4**
Illinois	0.0	0.0	87.6	16.5	0.0	0.0	0.0	87.6	16.5	104.2
Indiana	0.0	0.0	8.6	0.6	0.0	0.0	0.0	8.6	0.6	9.2
Iowa	0.0	0.0	1.7	0.5	0.0	0.0	0.0	1.7	0.5	2.2
Kentucky, Western	0.0	0.0	15.6	3.6	0.0	0.0	0.0	15.6	3.6	19.2
Missouri	0.0	0.0	1.5	4.5	0.0	0.0	0.0	1.5	4.5	6.0
Oklahoma	0.0	0.0	1.2	0.3	0.0	0.0	0.0	1.2	0.3	1.5
Texas	0.0	0.0	0.0	0.0	0.0	0.0	12.1	0.0	12.1	12.1
Other[c]	0.1	(s)	0.3	1.1	0.0	0.0	0.4	0.4	1.5	1.9
Western	**(s)**	**0.0**	**21.2**	**2.3**	**121.2**	**55.9**	**29.2**	**142.4**	**87.4**	**229.7**
Alaska	0.0	0.0	0.6	0.1	4.8	0.6	(s)	5.4	0.7	6.1
Colorado	(s)	0.0	7.5	0.6	3.7	0.0	4.2	11.2	4.8	15.9
Montana	0.0	0.0	1.4	0.0	69.6	32.3	15.8	70.9	48.0	119.0
New Mexico	(s)	0.0	2.7	0.9	3.4	5.0	0.0	6.1	5.9	12.0
North Dakota	0.0	0.0	0.0	0.0	0.0	0.0	8.9	0.0	8.9	8.9
Utah	0.0	0.0	4.9	0.3	(s)	0.0	0.0	4.9	0.3	5.2
Washington	0.0	0.0	0.3	0.0	1.0	0.0	(s)	1.3	(s)	1.3
Wyoming	0.0	0.0	3.8	0.5	38.6	18.1	0.0	42.5	18.5	61.0
Other[d]	0.0	0.0	(s)	0.0	(s)	(s)	0.4	(s)	0.4	0.4
U.S. total	**4.1**	**3.4**	**206.0**	**51.2**	**121.1**	**55.9**	**42.8**	**331.2**	**153.3**	**484.5**
States east of the Mississippi River	4.0	3.3	180.0	42.6	0.0	0.0	1.1	184.0	47.0	231.0
States west of the Mississippi River	0.1	(s)	25.9	8.6	121.1	55.9	41.7	147.2	106.3	253.5

[a]Lignite resources are not mined underground in the United States.
[b]Georgia, Maryland, North Carolina, and Tennessee.
[c]Arkansas, Kansas, Louisiana, and Michigan.
[d]Arizona, Idaho, Oregon, and South Dakota.
(s) = Less than 0.05 billion short tons.
Notes: Data represent remaining measured and indicated coal resources, analyzed and on file, meeting minimum seam and depth criteria, and in the ground as of January 1, 2011. These coal resources are not totally recoverable. Net recoverability with current mining technologies ranges from 0 percent (in far northern Alaska) to more than 90 percent. Fifty-four percent of the demonstrated reserve base of coal in the United States is estimated to be recoverable. Totals may not equal sum of components due to independent rounding.

SOURCE: "Table 4.8. Coal Demonstrated Reserve Base, January 1, 2011 (Billion Short Tons)," in *Annual Energy Review 2011*, U.S. Energy Information Administration, September 27, 2012, http://www.eia.gov/totalenergy/data/annual/pdf/aer.pdf (accessed October 3, 2012)

Nebraska, Nevada, North Dakota, Oregon, South Dakota, Texas, Utah, Virginia, and Washington.

From the late 1960s through the 1980s the domestic uranium industry provided most of the uranium that was used in U.S. nuclear power plants. (See Figure 5.4 in Chapter 5.) However, since that time purchased imports of uranium have far exceeded domestic production. In 2011 imported uranium totaled 54.4 million pounds (24.7 million kg), whereas domestically produced uranium totaled less than 4 million pounds (1.8 million kg). Despite the weak market for domestic uranium, exploration and drilling activities continued to occur. In 2011, 10,597 exploratory and development holes were drilled by the uranium mining industry. (See Table 7.4.) The total drilling footage in 2011 was 6.3 million feet (1.9 million m).

INTERNATIONAL RESERVES

When considering energy reserves outside of the United States, it is important to understand that many different terms are used internationally to refer to reserve volumes and to categorize them based on technological and economic criteria.

The EIA provides worldwide estimates of proved reserves of several fuels through its website "International Energy Statistics" (http://www.eia.gov/cfapps/ipdbproject/IEDIndex3.cfm). However, data are not available for all fuels, countries, or regions for every year.

Crude Oil and Natural Gas

According to the EIA, world proved reserves of crude oil totaled 1.3 trillion barrels in 2009. The countries with the

TABLE 7.4

Uranium drilling activities, 2003–11

Year	Exploration drilling		Development drilling		Exploration and development drilling	
	Number of holes	Feet (thousand)	Number of holes	Feet (thousand)	Number of holes	Feet (thousand)
2003	NA	NA	NA	NA	W	W
2004	W	W	W	W	2,185	1,249
2005	W	W	W	W	3,143	1,668
2006	1,473	821	3,430	1,892	4,903	2,713
2007	4,351	2,200	4,996	2,946	9,347	5,146
2008	5,198	2,543	4,157	2,551	9,355	5,093
2009	1,790	1,051	3,889	2,691	5,679	3,742
2010	2,439	1,460	4,770	3,444	7,209	4,904
2011	5,441	3,322	5,156	3,003	10,597	6,325

NA = Not available.
W = Data withheld to avoid disclosure of individual company data.
Note: Totals may not equal sum of components because of independent rounding.

SOURCE: "Table 1. U.S. Uranium Drilling Activities, 2003–2011," in *2011 Domestic Uranium Production Report*, U.S. Energy Information Administration, May 2012, http://www.eia.gov/uranium/production/annual/pdf/dupr.pdf (accessed October 16, 2012)

largest reserves were Saudi Arabia (267 billion barrels), Canada (178 billion barrels), Iran (136 billion barrels), Iraq (115 billion barrels), and Venezuela (99 billion barrels).

The EIA indicates that proved reserves of natural gas totaled 6,289 tcf (178.1 tcm) in 2009. The countries with the largest reserves were Russia (1,680 tcf [47.6 tcm]), Iran (992 tcf [28.1 tcm]), Qatar (892 tcf [25.3 tcm]), the United States (273 tcf [7.7 tcm]), and Saudi Arabia (258 tcf [7.3 tcm]).

Coal

The EIA reports in *International Energy Outlook 2011* (September 2011, http://www.eia.gov/forecasts/ieo/pdf/0484(2011).pdf) that worldwide TRR of coal totaled 948 billion tons (860 billion t) as of January 1, 2009. The largest reserves were held by the United States (260.6 billion tons [236.4 billion t]), Russia (173.1 billion tons [157 billion t]), China (126.2 billion tons [114.5 billion t]), Australia and New Zealand (84.8 billion tons [76.9 billion t]), and India (66.8 billion tons [60.6 billion t]).

Uranium

In "Supply of Uranium" (August 2012, http://www.world-nuclear.org/info/inf75.htm), the World Nuclear Association (WNA) notes that worldwide "known recoverable resources" of uranium totaled 5.9 million tons (5.3 million t) in 2011. According to the WNA, these resources included RAR and "inferred" resources, which are deposits "for which tonnage, grade and mineral content can be estimated with a low level of confidence." It should be noted that the WNA specifies that its listing of known recoverable resources for 2011 only includes deposits that could be recovered for up to $130 per 2.2 pounds (1 kg) as of January 11, 2011.

The countries with the largest "known recoverable resources" of uranium in 2011 were Australia (1.8 million tons [1.6 million t]), Kazakhstan (693,400 tons [629,000 t]), Russia (537,000 tons [487,200 t]), Canada (516,700 tons [468,700 t]), and Niger (464,100 tons [421,000 t]).

CHAPTER 8
ELECTRICITY

In 1879 Thomas Alva Edison (1847–1931) flipped the first switch to light Menlo Park, New Jersey. Since that time the use of electrical power has become nearly universal in the United States. Electricity is not really an energy source, but an energy carrier, because it is generated using primary energy sources, such as fossil fuels. As such, electricity supplies and pricing are highly dependent on the economic factors underlying the fuels and processes that are used to generate electricity. Likewise, the electric power sector is the focal point for environmental concerns that are associated with combusting or otherwise using primary energy sources.

UNDERSTANDING ELECTRICITY

Electricity results from the interaction of charged particles, such as electrons (negatively charged subatomic particles) and protons (positively charged subatomic particles). For example, static electricity is caused by friction: when one material rubs against another, it transfers charged particles. The zap people might feel and the spark people might see when they drag their feet on a carpet and then touch a metal doorknob demonstrate static electricity—electrons being transferred between a hand and the doorknob.

Generating Electricity

As noted in Chapter 1, electricity is not a primary energy source because energy is required to produce electricity. In the United States most electric power is produced by burning fossil fuels. The resulting heat turns water into steam, which can be used as a working fluid to turn the blades of a turbine. The hot gases from the burning fuels can also be used for this purpose. In either case, the turbine rotates a magnet that is nestled within or around coiled wire, which generates an electric current in the wire. Thus, heat is a primary factor in the operation of these plants. The same is true for nuclear power plants. They rely on the heat that is released from splitting uranium atoms apart to produce steam. (See Chapter 5.) Thermally derived electricity is also obtained from burning wood or waste, from geothermal reservoirs, and through some solar (sunlight-based) systems. These renewable energy sources are described in Chapter 6.

There are also nonthermal means for producing electricity. The most common is hydropower, which is described in Chapter 6. Hydropower relies on water (rather than on steam or hot gases) to be the working fluid that turns turbines. Wind is the working fluid that turns turbines at wind farms. Lastly, some solar systems turn sunlight directly into electrical current using crystalline materials.

Overall, the vast majority of electricity generation methods operated in the United States are thermally based—that is, they convert heat energy to mechanical energy to electrical energy. This conversion process is inherently inefficient because large amounts of energy are lost along the way.

Measuring Electricity

Electric current is the flow of electric charge; it is measured in amperes (amps). Electrical power is the rate at which energy is transferred by electric current. A watt is the standard measure of electrical power, named after the Scottish engineer James Watt (1736–1819). A watt is a very small unit of measure, so electrical power is typically measured in multiples of the watt, such as the kilowatt (kW; 1,000 watts), the megawatt (MW; 1 million watts), or the gigawatt (GW; 1 billion watts). Electrical work is usually measured using the kilowatt-hour (kWh), which is the work done by 1 kW acting for one hour. A 1,000-kW generator running at full capacity for one hour supplies 1,000 kWh of power. That generator operating continuously for an entire year produces nearly 8.8 million kWh of electricity (1,000 kW × 24 hours per day × 365 days per year).

Electric Power System/Electrical Grid

An electric power system, or electrical grid, is a network that connects the locations of electrical power generation with the locations of electrical power consumption. Figure 8.1 illustrates the several components of a simple electric system. Power plants generate electricity, which is raised to a high voltage for transmission. Voltage is a complicated concept, but basically raising the voltage improves the energy efficiency (i.e., reduces losses) as the transmission lines carry electricity over long distances. The voltage is stepped down (lowered) somewhat before the electricity goes onto the distribution lines that deliver it to neighborhoods, shopping centers, and other clusters of users. Then the voltage is stepped down again for delivery to individual homes and businesses. It should be noted that some industrial facilities can take their electricity at higher voltages than what is delivered to residential and commercial customers.

In the grid as a whole, substations connect the pieces of the system together, and energy control centers coordinate the operation of all the components.

DOMESTIC PRODUCTION AND CONSUMPTION

Electricity is produced on demand (i.e., as it is needed). However, consumer demand varies constantly as people turn on and off air conditioners, light switches, appliances, and other electrical devices. In "Factors Affecting Electricity Prices" (August 13, 2012, http://www.eia.gov/energyexplained/index.cfm?page=electricity_factors_affecting_prices), the Energy Information Administration (EIA), a division of the U.S. Department of Energy, indicates that electricity demand on a daily basis is usually highest in the afternoon and early evening. On a seasonal basis, demand is highest during the summer.

However, there are no large-scale means by which electricity can be efficiently stored for later use. This deficiency profoundly affects the electric power sector and the economics of electricity because it means the system as a whole has to be ready for peak demand at all times. In general, the power industry operates two kinds of generating capacity: baseload and peaking. The EIA explains in "Electric Generators' Roles Vary Due to Daily and Seasonal Variation in Demand" (June 8, 2011, http://www.eia.gov/todayinenergy/detail.cfm?id=1710) that "baseload capacity runs around the clock when it is not down for maintenance. Peaking capacity runs a few times a year for short periods to help electricity systems meet peak demand."

The vast majority of the nation's electricity is generated by the electric power sector; however, some industrial and commercial facilities operate their own power plants. Electricity produced by these facilities is called direct-use or self-generated electricity. The EIA notes in *Annual Energy Review 2011* (September 2012, http://www.eia.gov/totalenergy/data/annual/pdf/aer.pdf) that total electricity generation in 2011 was 4.1 trillion kWh. Of this total, 4 trillion kWh was produced by the electric power sector. These are net amounts—that is, they do not account for the electricity that was used by the power plants themselves.

The nation's electricity production has increased over the decades, but its growth has slowed recently. According to the EIA, in *Annual Energy Outlook 2012* (June 2012, http://www.eia.gov/forecasts/aeo/pdf/0383(2012).pdf), electricity demand grew by 9.8% annually between 1949 and 1959. By the 1970s the growth rate was roughly half that amount and continued to drop. During the first decade of the 21st century electricity demand grew by only 0.7% annually.

FIGURE 8.1

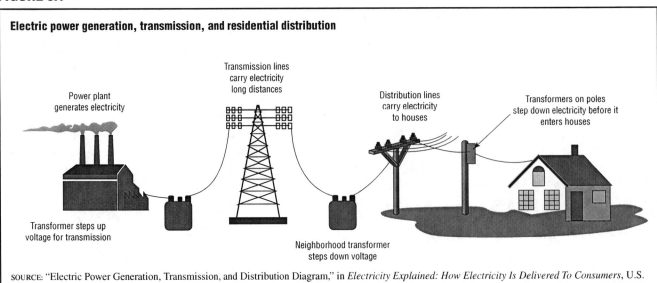

Electric power generation, transmission, and residential distribution

Transmission lines carry electricity long distances

Power plant generates electricity

Distribution lines carry electricity to houses

Transformers on poles step down electricity before it enters houses

Transformer steps up voltage for transmission

Neighborhood transformer steps down voltage

SOURCE: "Electric Power Generation, Transmission, and Distribution Diagram," in *Electricity Explained: How Electricity Is Delivered To Consumers*, U.S. Energy Information Administration, July 9, 2012, http://www.eia.gov/energyexplained/index.cfm?page=electricity_delivery (accessed October 16, 2012)

Production Sources

Figure 8.2 shows net electricity generation by source between 1949 and 2011. Coal has historically been the largest source. In 2011 it accounted for 42% of the total net generation of 4.1 trillion kWh. Natural gas was second (25%), followed by nuclear power (19%), hydropower (8%), and other sources (6%). The last category includes petroleum, which in decades past was widely used for electricity generation. Figure 8.3 traces the competitive forces that have influenced fossil fuel use by the electric power sector since 1950.

Coal has long been the fuel of choice because, as explained in Chapter 3, it is abundant domestically and relatively low-priced. Nuclear power began playing a larger role during the 1980s, but its use has been rather flat since the late 1990s. (See Figure 8.2.) The biggest movement in fuels has been by natural gas. During the first decade of the 21st century coal's supremacy was being seriously challenged by natural gas. This change was spurred by various technological, economic, and legislative factors.

According to the EIA, in "Most Electric Generating Capacity Additions in the Last Decade Were Natural Gas–Fired" (July 5, 2011, http://www.eia.gov/todayin energy/detail.cfm?id=2070), more than 80% of total generation capacity additions made between 2000 and 2010 were natural gas–fired. This growth reflects market changes that began during the late 1980s:

- Decreasing natural gas prices and increasing domestic supplies.

- Increasing availability of combined-cycle units, which are more efficient than steam-only units.

As noted earlier, the working fluid that turns a turbine can be hot gases or steam. In a combined-cycle plant, both types of turbines are used, in that hot gases leaving one turbine are used to heat water to produce steam to turn another turbine.

- Repeal of provisions contained in a 1978 law that "discouraged" the use of natural gas for electricity generation.

In addition, natural gas became a favored fuel for power generation because it was cheaper and faster to construct natural gas–fired plants than coal-fired plants.

It is important to understand that fossil fuel–fired power plants do not necessarily use a single fuel for all of their capacity. For example, a plant might use natural gas for its baseload capacity and petroleum for its peaking capacity. In addition, power companies typically operate multiple power plants. Thus, the companies can concentrate their capacity at different times on different fuels, which is a process called fuel switching. In "Natural Gas Consumption Reflects Shifting Sectoral Patterns" (May 16, 2012, http://www.eia.gov/todayinenergy/detail.cfm?id=6290), the EIA explains that "the electric power sector has the flexibility to shift some amount of baseload power generation, much of which has traditionally been fueled by coal, to underutilized natural gas generators without requiring additional investments in infrastructure."

Consumption

Figure 8.4 shows retail electricity sales for various economic sectors. The residential sector was the largest

FIGURE 8.2

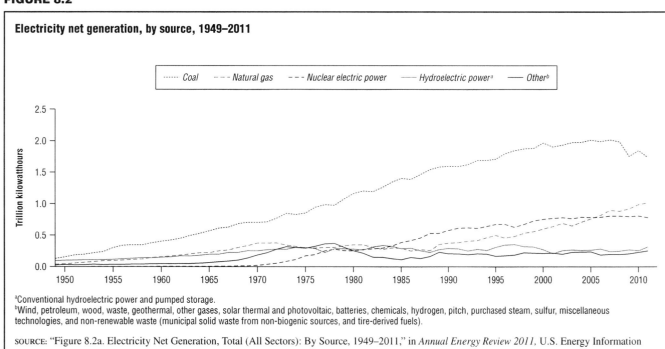

Electricity net generation, by source, 1949–2011

aConventional hydroelectric power and pumped storage.
bWind, petroleum, wood, waste, geothermal, other gases, solar thermal and photovoltaic, batteries, chemicals, hydrogen, pitch, purchased steam, sulfur, miscellaneous technologies, and non-renewable waste (municipal solid waste from non-biogenic sources, and tire-derived fuels).

SOURCE: "Figure 8.2a. Electricity Net Generation, Total (All Sectors): By Source, 1949–2011," in *Annual Energy Review 2011*, U.S. Energy Information Administration, September 27, 2012, http://www.eia.gov/totalenergy/data/annual/pdf/aer.pdf (accessed October 3, 2012)

FIGURE 8.3

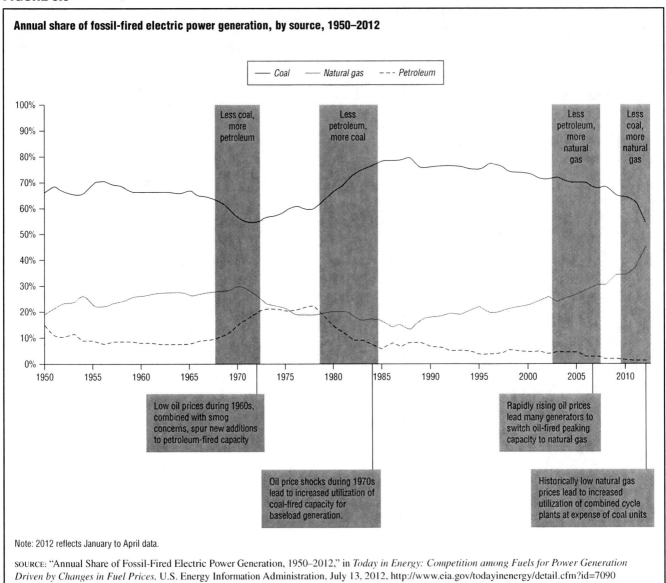

Annual share of fossil-fired electric power generation, by source, 1950–2012

— Coal   — Natural gas   - - - Petroleum

Less coal, more petroleum

Less petroleum, more coal

Less petroleum, more natural gas

Less coal, more natural gas

Low oil prices during 1960s, combined with smog concerns, spur new additions to petroleum-fired capacity

Oil price shocks during 1970s lead to increased utilization of coal-fired capacity for baseload generation.

Rapidly rising oil prices lead many generators to switch oil-fired peaking capacity to natural gas

Historically low natural gas prices lead to increased utilization of combined cycle plants at expense of coal units

Note: 2012 reflects January to April data.

SOURCE: "Annual Share of Fossil-Fired Electric Power Generation, 1950–2012," in *Today in Energy: Competition among Fuels for Power Generation Driven by Changes in Fuel Prices*, U.S. Energy Information Administration, July 13, 2012, http://www.eia.gov/todayinenergy/detail.cfm?id=7090 (accessed October 17, 2012)

user in 2011 at 1.4 trillion kWh. It was followed closely by the commercial sector, at 1.3 trillion kWh, and the industrial sector, at 1 trillion kWh. The transportation sector was a very small electricity consumer in 2011, accounting for only 0.01 trillion kWh.

From 1949 to the mid-1980s the industrial sector was the largest consumer of electricity in the United States, and its usage rate grew quickly. (See Figure 8.4.) That growth slowed through the 1990s, and during the first decade of the 21st century industrial consumption was flat to slightly declining. Residential and commercial use grew steadily until around 2005 and then flattened out somewhat.

Losses

As noted earlier, energy is lost when thermal energy is converted to mechanical energy and then to electrical

energy. Conversion losses are significant. Figure 8.5 shows an energy balance for electricity flows in 2011. Just over 40 quadrillion British thermal units (Btu) of energy was input to electricity generation; however, 63% (25.2 quadrillion Btu) of it was lost during conversion, leaving 14.8 quadrillion Btu of gross generation. An additional 1 quadrillion Btu was lost during transmission and distribution of the net generated electricity. Overall, just over 26 quadrillion Btu, or two-thirds of the incoming energy, was lost.

ELECTRICITY PRICES

Figure 8.6 shows the average retail prices of electricity by sector between 1960 and 2011. These prices are expressed in real (inflation-adjusted) cents. Real price amounts assume that the value of a dollar is constant over time. Thus, changes reflect actual market variations

FIGURE 8.4

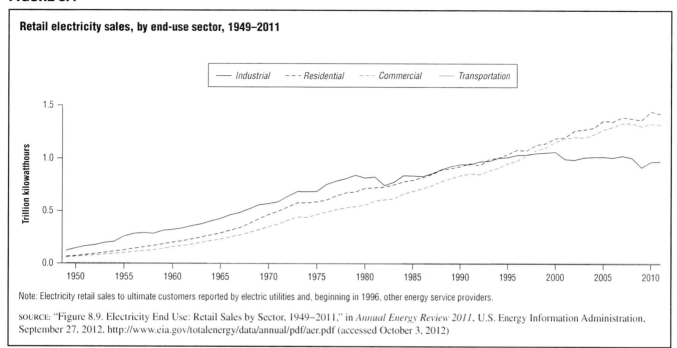

Retail electricity sales, by end-use sector, 1949–2011

— Industrial   - - - Residential   ···· Commercial   ····· Transportation

Note: Electricity retail sales to ultimate customers reported by electric utilities and, beginning in 1996, other energy service providers.

SOURCE: "Figure 8.9. Electricity End Use: Retail Sales by Sector, 1949–2011," in *Annual Energy Review 2011*, U.S. Energy Information Administration, September 27, 2012, http://www.eia.gov/totalenergy/data/annual/pdf/aer.pdf (accessed October 3, 2012)

and not inflationary effects. Retail prices were quite variable by sector during the 1960s, but prices for all sectors dipped through the early 1970s and then spiked back up through the early 1980s. Prices dropped through the end of the century and then began creeping upward again.

Pricing Factors

In "Factors Affecting Electricity Prices," the EIA notes that electricity prices are affected by fuel costs, power plant construction and maintenance costs, and transmission and distribution line costs. Weather also plays a factor. As noted earlier, electricity usage is higher during the summer than during other seasons. Hot summer weather can greatly increase the demand for electricity for air conditioning and drive up prices. Lastly, electricity prices are affected by government regulations and policies, which will be explained later in this chapter.

The EIA estimates that the average electricity price in the United States is affected mostly by generation costs (58% contribution) and then by distribution costs (31% contribution) and transmission costs (11% contribution). (See Figure 8.7.)

The prices that are paid by electricity consumers vary by economic sector. According to the EIA, in "Factors Affecting Electricity Prices," the average retail price of electricity sold in the United States in 2011 was 10 cents per kWh. The residential sector paid the highest price (11.8 cents per kWh), followed by the transportation sector (10.6 cents per kWh), the commercial sector (10.3 cents per kWh), and the industrial sector (6.9 cents per kWh). The EIA notes that residential and commercial customers pay the most because delivery to them is the most expensive. By contrast, industrial customers pay the least because they purchase electricity in large amounts and because many industrial facilities do not require their electricity to be stepped down by transformers—that is, their systems can handle electricity at higher voltages than what is delivered to commercial and residential customers.

The EIA indicates that electricity prices vary considerably state by state. For example, in 2010 the average price was only 6.2 cents per kWh in Wyoming, but 25.1 cents per kWh in Hawaii. (An EIA listing of electricity prices by state is available at http://www.eia.gov/electricity/state/.) Pricing differences between states are due to various factors, including the number and type of power plants in an area, the availability and price of fuel in local markets, and government regulations and policies that affect pricing.

GOVERNMENT INTERVENTION. Electricity is considered to be a vital resource for the public. Hence, for decades the government heavily controlled the nation's electricity markets to ensure that this resource was widely available at relatively low costs. As a result, local utility companies developed monopolies in that each company had market control over a specified geographical area. During the 1970s and 1980s the federal government began experimenting with deregulation (breaking up government-supported monopolies). The idea was that competition would keep electricity prices low.

In the electric power sector deregulation has meant allowing consumers to purchase electricity from competing sellers, but still receive delivery over existing power lines that are maintained by a local utility. The overall

FIGURE 8.5

Electricity flow, 2011

[Quadrillion Btu]

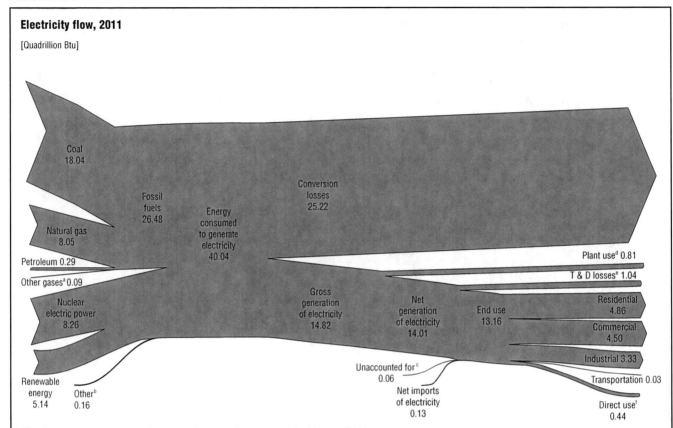

[a]Blast furnace gas, propane gas, and other manufactured and waste gases derived from fossil fuels.
[b]Batteries, chemicals, hydrogen, pitch, purchased steam, sulfur, miscellaneous technologies, and non-renewable waste (municipal solid waste from non-biogenic sources, and tire-derived fuels).
[c]Data collection frame differences and nonsampling error. Derived for the diagram by subtracting the "T & D Losses" estimate from "T & D Losses and Unaccounted for."
[d]Electric energy used in the operation of power plants.
[e]Transmission and distribution losses (electricity losses that occur between the point of generation and delivery to the customer) are estimated as 7 percent of gross generation.
[f]Use of electricity that is 1) self-generated, 2) produced by either the same entity that consumes the power or an affiliate, and 3) used in direct support of a service or industrial process located within the same facility or group of facilities that house the generating equipment. Direct use is exclusive of station use.
Notes: Data are preliminary. Net generation of electricity includes pumped storage facility production minus energy used for pumping. Values are derived from source data prior to rounding for publication. Totals may not equal sum of components due to independent rounding.
Btu = British thermal unit.

SOURCE: "Figure 8.0. Electricity Flow, 2011 (Quadrillion Btu)," in *Annual Energy Review 2011*, U.S. Energy Information Administration, September 27, 2012, http://www.eia.gov/totalenergy/data/annual/pdf/aer.pdf (accessed October 3, 2012)

process is known as restructuring. In 1978 Congress passed the Public Utilities Regulatory Policies Act, which required utilities to buy electricity from private companies when doing so was cheaper than building their own power plants. The Energy Policy Act of 1992 gave other electricity generators greater access to the market, which enhanced the states' ability to restructure their systems. Widespread debates occurred regarding regulatory, economic, energy, and environmental policies. State public utility commissions conducted proceedings and crafted rules that were related to competition.

California was a leader in deregulation during the mid-1990s. However, during the summer of 2000 the state experienced rolling electrical blackouts, and electricity bills doubled for many customers. Fearful of similar blackouts and price spikes, most other states slowed or stopped their efforts to deregulate their electricity markets. As a result, as of December 2012 few states

had deregulated their electricity markets to any degree. Northeastern states had most fully deregulated, and Texas had competitive markets in most, but not all, of its major cities. However, it is difficult to determine how deregulation has affected electricity pricing because so many different variables are at work.

Some electricity providers are government entities, such as municipally owned utilities. In addition, as noted in Chapter 6, the federal government generates electricity for consumers through federally owned corporations, such as the Tennessee Valley Authority.

The federal government also supports the electric power sector via financial incentives, such as tax breaks and subsidies. In *Direct Federal Financial Interventions and Subsidies in Energy in Fiscal Year 2010* (July 2011, http://www.eia.gov/analysis/requests/subsidy/pdf/subsidy .pdf), the EIA indicates that the electric power sector

FIGURE 8.6

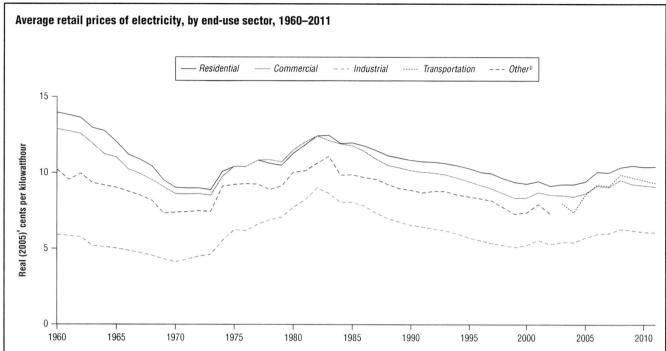

Average retail prices of electricity, by end-use sector, 1960–2011

— Residential ······ Commercial – – – Industrial ······· Transportation – – – Other[b]

[a]In chained (2005) dollars, calculated by using gross domestic product implicit price deflators.
[b]Public street and highway lighting, interdepartmental sales, other sales to public authorities, agriculture and irrigation, and transportation including railroads and railways.
Note: Taxes are included.

SOURCE: "Figure 8.10. Average Retail Prices of Electricity: By Sector, Real Prices, 1960–2011," in *Annual Energy Review 2011*, U.S. Energy Information Administration, September 27, 2012, http://www.eia.gov/totalenergy/data/annual/pdf/aer.pdf (accessed October 3, 2012)

FIGURE 8.7

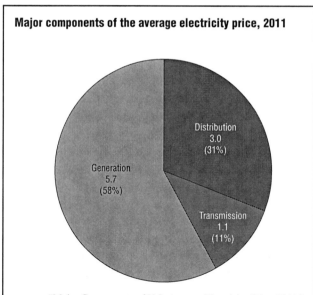

Major components of the average electricity price, 2011

Distribution
3.0
(31%)

Generation
5.7
(58%)

Transmission
1.1
(11%)

SOURCE: "Major Components of U.S. Average Electricity Price, 2011," in *Electricity Explained: Factors Affecting Electricity Prices*, U.S. Energy Information Administration, August 13, 2012, http://www.eia.gov/energyexplained/index.cfm?page=electricity_factors_affecting_prices (accessed October 16, 2012)

received $971 million in direct expenditures, tax breaks, research and development funds, loan guarantees, and other means of financial support in 2010. This was around 3% of the $37.2 billion that was allocated to the energy industry as a whole. The funds provided in 2010 were designed to facilitate the nation's smart grid and transmission technologies. Smart grid technologies are technologies that are designed to make the nation's electrical grid "smarter" (i.e., more computerized and responsive). According to the EIA (April 2009, http://www.eia.gov/oiaf/servicerpt/stimulus/arra.html), these technologies include "a wide array of measurement, communications, and control equipment employed throughout the transmission and distribution system that will enable real-time monitoring of the production, flow, and use of power from generator to consumer." The agency notes that full deployment of the technologies should improve the grid's efficiency, lower costs, allow greater use of renewable energy sources for power generation, and provide more information to providers and consumers about electricity consumption.

Whereas government subsidies and tax breaks put downward pressure on electricity prices, other government activities tend to push prices upward. The primary example is the regulation of emissions and discharges from fuel combustion. As noted earlier, the electric power sector is the biggest end-user of coal in the United States. However, coal-fired power plants have been under increasing pressure to reduce their emissions of air pollutants. Stricter standards have forced coal-fired power plants to invest in better control technologies and/or use coal that contains lower sulfur contents. Pollution-control technologies, in particular, can be very expensive.

THE DOMESTIC OUTLOOK

In *Annual Energy Outlook 2012*, the EIA predicts future domestic electricity demand for three cases of economic growth. (See Figure 8.8.) Overall, electricity demand is projected to grow slowly through 2035. The projected growth rate is around 1% for the reference case. The agency forecasts this low growth rate because it believes that increasing demand for electricity from a growing population will be offset by energy efficiency improvements in appliances and other electricity-using equipment.

The EIA anticipates that coal will continue to be the largest energy source for electricity generation through 2035; however, its share of total consumption is expected to shrink from 45% in 2010 to 38% in 2035. (See Figure 8.9.) By contrast, natural gas's share is projected to increase from 24% in 2010 to 28% in 2035. Thus, fossil fuels are forecast to account for 66% of electricity generation in 2035. Nuclear power is projected to account for 18% of the total in 2035 (down from 20% in 2010), while renewables, mostly wind power and biomass combustion, will supply 15% of the total (up from 10% in 2010).

FIGURE 8.8

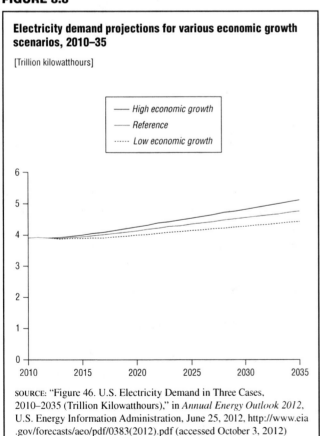

Electricity demand projections for various economic growth scenarios, 2010–35

[Trillion kilowatthours]

—— High economic growth
— Reference
····· Low economic growth

SOURCE: "Figure 46. U.S. Electricity Demand in Three Cases, 2010–2035 (Trillion Kilowatthours)," in *Annual Energy Outlook 2012*, U.S. Energy Information Administration, June 25, 2012, http://www.eia.gov/forecasts/aeo/pdf/0383(2012).pdf (accessed October 3, 2012)

FIGURE 8.9

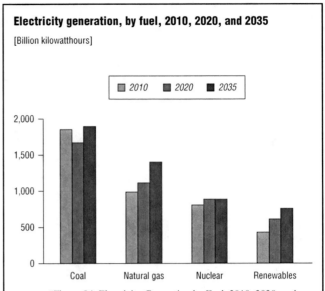

Electricity generation, by fuel, 2010, 2020, and 2035

[Billion kilowatthours]

2010 2020 2035

SOURCE: "Figure 94. Electricity Generation by Fuel, 2010, 2020, and 2035 (Billion Kilowatthours)," in *Annual Energy Outlook 2012*, U.S. Energy Information Administration, June 25, 2012, http://www.eia.gov/forecasts/aeo/pdf/0383(2012).pdf (accessed October 3, 2012)

WORLD ELECTRICITY PRODUCTION AND CONSUMPTION

Net electricity generation worldwide totaled 19.1 trillion kWh in 2008. (See Table 8.1.) Coal provided 7.7 trillion kWh, or 40% of the total. It was followed by natural gas, at 4.2 trillion kWh (22% of total); renewable sources, at 3.7 trillion kWh (19% of total); nuclear energy, at 2.6 trillion kWh (14% of total); and liquids, such as petroleum, at 1 trillion kWh (5% of the total).

The EIA examines and predicts electricity generation for countries that are and are not members of the Organisation for Economic Co-operation and Development (OECD). As explained in Chapter 5, the OECD is a collection of dozens of mostly western nations that are devoted to global economic development. The total net electricity generation in 2008 was split roughly evenly between OECD members and nonmembers. (See Table 8.1 and Figure 8.10.) OECD members showed greater generation than nonmembers from natural gas and nuclear power, whereas nonmembers had greater generation than OECD members from liquid fuels, coal, and renewables.

OECD members are expected to make small gains in electricity generation from 10.2 trillion kWh in 2008 to 13.9 trillion kWh in 2035, for an annual growth rate of 1.2%. Most of this growth will be from renewables and natural gas. Nonmember electricity generation is projected to grow from 8.9 trillion kWh in 2008 to 21.2 trillion kWh in 2035, for an annual growth rate of 3.3%. Renewables, natural gas, and nuclear power are expected to fuel most of the capacity additions.

TABLE 8.1

World net electricity generation, by country category and energy source, 2008 and 2035

[Trillion kilowatthours]

Region	2008	2035	Average annual percent change, 2008–2035
OECD			
Liquids	0.4	0.3	−0.8
Natural gas	2.3	3.8	1.8
Coal	3.6	3.8	0.2
Nuclear	2.2	2.9	1.0
Renewables	1.8	3.2	2.2
Total OECD	**10.2**	**13.9**	**1.2**
Non-OECD			
Liquids	0.7	0.5	−1.0
Natural gas	1.8	4.6	3.4
Coal	4.1	9.1	3.0
Nuclear	0.4	2.0	6.0
Renewables	1.9	5.0	3.7
Total non-OECD	**8.9**	**21.2**	**3.3**
World			
Liquids	1.0	0.8	−0.9
Natural gas	4.2	8.4	2.6
Coal	7.7	12.9	1.9
Nuclear	2.6	4.9	2.4
Renewables	3.7	8.2	3.1
Total World	**19.1**	**35.2**	**2.3**

Note: Totals may not equal sum of components due to independent rounding.

SOURCE: Adapted from "Table 11. OECD and Non-OECD Net Electricity Generation by Energy Source, 2008–2035 (Trillion Kilowatthours)," in *International Energy Outlook 2011*, U.S. Energy Information Administration, September 19, 2011, http://www.eia.gov/forecasts/ieo/pdf/0484(2011).pdf (accessed October 3, 2012)

FIGURE 8.10

World net electricity generation, by country category, 1990–2035

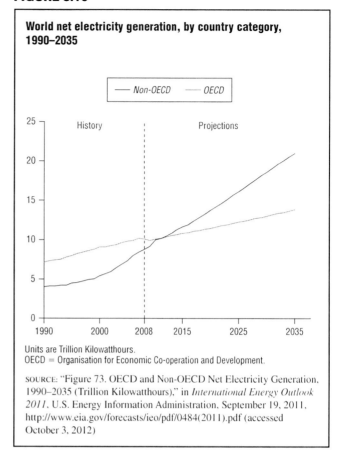

Units are Trillion Kilowatthours.
OECD = Organisation for Economic Co-operation and Development.

SOURCE: "Figure 73. OECD and Non-OECD Net Electricity Generation, 1990–2035 (Trillion Kilowatthours)," in *International Energy Outlook 2011*, U.S. Energy Information Administration, September 19, 2011, http://www.eia.gov/forecasts/ieo/pdf/0484(2011).pdf (accessed October 3, 2012)

ENVIRONMENTAL ISSUES

The primary environmental issues involved with electricity generation relate to the underlying fuels and methods that are used to generate power. These issues are briefly described for natural gas in Chapter 3, for coal in Chapter 4, for nuclear power in Chapter 5, and for hydropower and renewables in Chapter 6.

The electric power sector is heavily regulated for environmental compliance, particularly in regards to air emissions from the combustion of fossil fuels. Historically, these regulations have not covered carbon dioxide, which is a major contributor to global warming and associated climate change. However, as noted in Chapter 4, in March 2012 the U.S. Environmental Protection Agency (http://epa.gov/carbonpollutionstandard/actions.html) proposed limiting carbon emissions from newly constructed power plants. As of December 2012, the regulations had not been implemented. However, emerging technologies related to carbon capture and storage were gaining importance as a means for mitigating carbon buildup in the atmosphere. Rather than limiting emissions by using control equipment, these technologies focus on capturing carbon emissions (e.g., in the form of carbon dioxide) and either selling them or sequestering them. Sequestration involves long-term storage in repositories, such as geological formations deep underground.

The Department of Energy indicates in "DOE-Sponsored Project Begins Demonstrating CCUS Technology in Alabama" (http://www.fossil.energy.gov/news/techlines/2012/12037-CO2_Injection_Begins_in_Alabama.html) that in August 2012 "the world's first fully integrated coal power and geologic storage project" began injecting carbon dioxide into a geological repository in Alabama. The emissions are from Alabama Power's Plant Barry; the captured carbon dioxide is piped approximately 12 miles (19 km) away for injection into a geological formation more than 9,000 feet (2,700 m) underground. As of December 2012, other carbon capture projects were being developed around the country and worldwide.

CHAPTER 9
ENERGY CONSERVATION

The word *conservation* has different meanings depending on the context in which it is used. When referring to a resource, such as energy, to conserve means to use less of the resource. Of course, energy is a vital resource to a modern society with a growing population and a growing economy. It is a difficult proposition to use less energy and keep growing. One key to achieving this goal is to increase energy efficiency (i.e., decrease the amount of energy required to do a certain amount of work). Ever since the start of the Industrial Revolution (1760–1848), innovators have greatly enhanced the efficiency of energy-using vehicles, machines, and processes. Some of these innovations have been driven solely by consumer demand, whereas others have been motivated by government mandates. As will be explained in this chapter, the government plays a major role in pushing energy conservation to further specific goals.

Consumers practice energy conservation in accordance with their budgets. When energy prices rise to an uncomfortable level, consumers cut back their energy spending. The United States uses a mix of energy sources for which prices can vary considerably. In theory, energy consumers can switch from more expensive sources to cheaper sources as prices change. This is difficult in practice, however, because energy-consuming machines tend to be sole-source (e.g., most cars run only on gasoline). Even though there is some flexibility for fuel switching in the industrial and electric power sectors, residential and commercial consumers have fewer options in this regard. Nevertheless, they can choose from a variety of efficiency choices within product lines, such as cars, furnaces, air conditioning systems, and even lightbulbs.

Domestic energy self-sufficiency (or energy security) has long been a national goal for the United States. Even though domestic production of most energy sources meets domestic demand, the glaring exception is oil. As explained in Chapter 2, the United States imports large amounts of petroleum each year from the Middle East, a region that is riddled with political insecurity and a history of poor relations with the United States. Since the 1970s the United States has greatly reduced its reliance on oil imports from the Middle East. This has been achieved, in part, through persistent government focus on reducing overall petroleum usage by U.S. consumers.

Another motivator for energy conservation lies in environmental concerns about energy sources. The U.S. government enforces numerous regulations that are designed to prevent, reduce, and mitigate the worst environmental impacts from energy production and usage. These measures have had varying levels of success. Historically, the government has focused on air pollutants, such as sulfur dioxide, that result from the combustion of fossil fuels and are known to damage ecosystems and human health. During the 20th century the world became aware of global warming. Scientists believe that the release of large amounts of carbon into the atmosphere has caused the earth's atmosphere to warm unnaturally. (Carbon-containing gases are also known as greenhouse gases.) Anthropogenic (human-caused) warming is occurring at an alarming rate and is precipitating climate and ecosystem changes around the world. Concern about global warming has become a new, and powerful, impetus for reducing fossil fuel combustion.

NATIONAL ENERGY CONSERVATION

At first glance, it appears that the United States has not been successful at conserving energy. Domestic energy consumption increased from 32 quadrillion British thermal units (Btu) in 1949 to 97.3 quadrillion Btu in 2011, an increase of 204%. (See Figure 1.3 in Chapter 1.) However, this trend is not unexpected, given that the U.S. population and economy grew tremendously during this period.

National energy efficiency can be measured using two indicators. The first is energy consumption per capita (per person). Energy consumption per capita was 214 million Btu in 1949. (See Figure 9.1.) In 2011 it was 312 million Btu, a 46% increase compared with 1949. Thus, energy consumption per capita grew much slower than did total energy consumption. According to the Energy Information Administration (EIA) within the U.S. Department of Energy (DOE), in *Annual Energy Review 2011* (September 2012, http://www.eia.gov/total energy/data/annual/pdf/aer.pdf), in 2011 the U.S. population was 311.6 million. If energy consumption per capita had grown by 204% between 1949 and 2011 (as did total consumption), then energy consumption per capita in 2011 would have been 651 million Btu, more than twice its actual value.

A second indicator of efficiency is energy consumption per dollar of gross domestic product (GDP; the total value of goods and services that are produced by a nation). The GDP is a measure of national economic well-being; a growing GDP over time indicates a growing and thriving economy. The EIA indicates that the U.S. GDP grew from $1.8 trillion in 1949 to $13.3 trillion in 2011, an increase of 622%. (Note that these values are expressed in real [inflation-adjusted] dollars, which assume that a dollar had the same value during the entire period.) Energy consumption per real dollar of GDP declined from 17.4 thousand Btu in 1949 to 7.3 thousand Btu in 2011, a 58% decrease. (See Figure 9.2.)

FIGURE 9.1

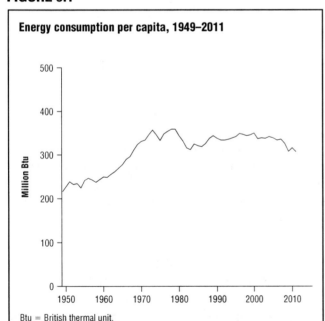

Energy consumption per capita, 1949–2011

Btu = British thermal unit.

SOURCE: "Figure 1.5. Energy Consumption and Expenditures Indicators Estimates: Energy Consumption per Capita, 1949–2011," in *Annual Energy Review 2011*, U.S. Energy Information Administration, September 27, 2012, http://www.eia.gov/totalenergy/data/annual/pdf/aer.pdf (accessed October 3, 2012)

FIGURE 9.2

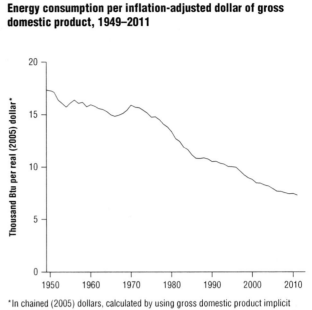

Energy consumption per inflation-adjusted dollar of gross domestic product, 1949–2011

*In chained (2005) dollars, calculated by using gross domestic product implicit price deflators.
Btu = British thermal unit.

SOURCE: "Figure 1.5. Energy Consumption and Expenditures Indicators Estimates: Energy Consumption per Real Dollar of Gross Domestic Product, 1949–2011," in *Annual Energy Review 2011*, U.S. Energy Information Administration, September 27, 2012, http://www.eia.gov/totalenergy/data/annual/pdf/aer.pdf (accessed October 3, 2012)

To summarize, total domestic energy consumption grew by 204% between 1949 and 2011, whereas energy consumption per capita grew by only 46% and energy consumption per real dollar of GDP declined by 58%. Clearly, the United States improved its energy efficiency to a great extent during this period. These gains were achieved to varying degrees across all sectors: electric power, transportation, industrial, and residential and commercial.

THE ELECTRIC POWER SECTOR

The electric power sector was the largest consumer of primary energy in 2011, accounting for 40% of the total. (See Figure 1.10 in Chapter 1.) Figure 8.2 in Chapter 8 indicates that the sector used fossil fuels for 67% of electricity generation that year (42% from coal and 25% from natural gas). Nuclear energy accounted for another 19%, hydroelectric power for 8%, and other sources for 6% of the total. In *Annual Energy Review 2011*, the EIA notes that net electricity generation increased from 2,967.1 billion kilowatt-hours (kWh) in 1989 to 4,105.7 billion kWh in 2011. Electricity demand has been driven by growing consumption in the residential, commercial, and industrial sectors.

The electric power sector experiences huge energy losses between inputs and outputs. As explained in Chapter 8, around two-thirds of the incoming energy is lost during generation. The sector has focused its conservation efforts on enhancing the efficiency of individual power plants and their systems. One innovation is the combined-cycle power plant, which reuses the heat leaving one turbine to boil water into steam to turn another turbine. Efficiency gains such as this have allowed the industry to generate more power using less fuel. In addition, the sector as a whole has enhanced its efficiency over time by retiring older less-efficient plants and equipment in favor of newer models.

THE TRANSPORTATION SECTOR

The U.S. transportation system plays a central role in the economy and is a major energy consumer. Figure 1.10 in Chapter 1 shows that the transportation sector accounted for 28% of total primary energy consumption in 2011. As explained in Chapter 1, this sector's energy consumption is for vehicles whose primary purpose is transporting people and/or goods from place to place. Transportation vehicles include automobiles; trucks; buses; motorcycles; trains, subways, and other rail vehicles; aircraft; and ships, barges, and other waterborne vehicles. Americans used 27 quadrillion Btu of energy for transportation in 2011, of which petroleum made up 93% of the total. In *Annual Energy Outlook 2012* (June 2012, http://www.eia.gov/forecasts/aeo/pdf/0383(2012).pdf), the EIA estimates the relative consumption contributions for various transport modes. Highway vehicles are divided into two categories: light duty (e.g., cars, sport-utility vehicles, and minivans) and heavy duty (e.g., tractor trailers, buses, and delivery vans). In 2010 light-duty vehicles accounted for nearly two-thirds (62%) of total transportation energy use, followed by heavy-duty vehicles (18%). Air, marine, and rail vehicles were minor energy consumers.

Historical petroleum consumption by sector is shown in Figure 2.10 in Chapter 2. Between 1973 and 2011 petroleum usage by the industrial, residential and commercial, and electric power sectors was flat to declining, which was achieved through a combination of fuel switching and efficiency gains. In contrast, petroleum consumption by the transportation sector grew from 9.1 million barrels per day in 1973 to 13.3 million barrels per day in 2011, a 46% increase. In part, this growth was due to an increasing population and more vehicles on the road. In addition, vehicle types changed dramatically during this period. The DOE's Oak Ridge National Laboratory (ORNL) indicates in *Transportation Energy Data Book, Edition 31* (July 2012, http://cta.ornl.gov/data/tedb31/Edition31_Full_Doc.pdf) that in 1975 cars accounted for more than 70% of total sales of light-duty vehicles. By 2011 that percentage had dropped to 49% as sport-utility vehicles, minivans, and pickup trucks dominated the market. These larger vehicles have inherently poorer fuel economy than cars. (See Figure 9.3.) As explained in Chapter 2, oil prices were historically low

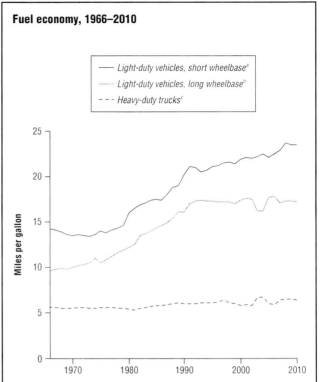

FIGURE 9.3

Fuel economy, 1966–2010

— Light-duty vehicles, short wheelbase[a]
...... Light-duty vehicles, long wheelbase[b]
- - - Heavy-duty trucks[c]

[a]Through 2006, data are for passenger cars (and, through 1989, for motorcycles). Beginning in 2007, data are for passenger cars, light trucks, vans, and sport utility vehicles with a wheel-base equal to or less than 121 inches.
[b]Through 2006, data are for vans, pickup trucks, sport utility vehicles, and a small number of trucks with 2 axles and 4 tires, such as step vans. Beginning in 2007, data are for large passenger cars, vans, pickup trucks, and sport utility vehicles with a wheelbase larger than 121 inches.
[c]Through 2006, data are for single-unit trucks with 2 axles and 6 or more tires, and combination trucks. Beginning in 2007, data are for single-unit trucks with 2 axles and 6 or more tires or a gross vehicle weight rating exceeding 10,000 pounds, and combination trucks.

SOURCE: "Figure 2.8. Motor Vehicle Mileage, Fuel Consumption, and Fuel Economy: Fuel Economy, 1966–2010," in *Annual Energy Review 2011*, U.S. Energy Information Administration, September 27, 2012, http://www.eia.gov/totalenergy/data/annual/pdf/aer.pdf (accessed October 3, 2012)

from the mid-1980s through the end of the 1990s. During this period Americans bought and drove larger vehicles to satisfy their personal preferences. This highlights the huge influence that market prices have on consumer demand and shows how difficult it can be to get consumers to conserve oil by their own accord.

Government Intervention

Chapter 2 describes the energy crisis (actually an oil crisis) that stunned the nation during the early 1970s. The political and economic consequences of the United States' dependence on foreign oil spurred government actions to reduce domestic oil consumption, particularly in the transportation sector. In *Saving Energy in U.S. Transportation* (July 1994, http://govinfo.library.unt.edu/ota/Ota_1/DATA/1994/9432.PDF), the Office of Technology Assessment lists the types of conservation measures that government entities have used over the years:

- Require vehicles to have higher fuel economy
- Design highways to optimize traffic flow and reduce fuel consumption
- Impose higher taxes on petroleum fuels and/or low-fuel-economy vehicles
- Use tax credits and subsidies to support research and development devoted to petroleum conservation
- Pass regulations requiring stricter inspection and maintenance programs for vehicles
- Encourage carpooling and working from home
- Add and improve mass transit options
- Mandate the use of alternative-fuel (nonpetroleum) vehicles and alternative fuels
- Offer tax credits and other financial incentives to encourage petroleum conservation by consumers

Not all the measures are mutually supportive. For example, efforts to promote a freer flow of automobile traffic, such as high-occupancy vehicle lanes or free parking for carpools, may sabotage efforts to shift travelers to mass transit or reduce trip lengths and frequency.

FUEL ECONOMY STANDARDS. Of all the measures that the federal government uses to enhance oil conservation, the most well-known and perhaps most controversial are fuel economy standards. These standards represent a high level of government intervention in private markets. In 1975 the Energy Policy and Conservation Act set the initial Corporate Average Fuel Economy (CAFE) standards; since then, the standards have been modified numerous times.

The first CAFE standards required domestic automakers to increase the average mileage of new cars sold to 27.5 miles per gallon (mpg; 8.6 L/100 km) by 1985. Manufacturers could still sell large, less-efficient cars, but to meet the average fuel efficiency rates, they also had to sell smaller, more efficient cars. Automakers that failed to meet each year's standards were fined; those that managed to surpass the rates earned credits they could use in years when they fell below the requirements. Even while keeping some models relatively large and roomy, the manufacturers managed to improve mileage with innovations such as electronic fuel injection, which supplied fuel to an automotive engine more efficiently than its predecessor, the carburetor.

The EIA indicates in *Annual Energy Outlook 2012* that the nation's fuel economy for light-duty vehicles increased from an average of 19.9 mpg in 1978 to 26.2 mpg in 1987. Over the following two decades, however, the average fell to 24 to 26 mpg (9.8 to 9 L/100 km) due to an increase of larger light-duty vehicles on the market. New CAFE standards that applied to model year (MY) 2008 light-duty trucks helped push the national average fuel economy up to 29.2 mpg in 2010. These fuel standards were the first to be attribute-based, meaning that they varied depending on a specific vehicle attribute, in this case the vehicle's footprint (i.e., the surface area between all of its wheels). Thus, attribute-based CAFE standards vary by vehicle model.

In the press release "Obama Administration Finalizes Historic 54.5 mpg Fuel Efficiency Standards" (August 28, 2012, http://www.nhtsa.gov/), the National Highway Traffic Safety Administration (NHTSA) notes that in August 2012 the administration of President Barack Obama (1961–) finalized fuel economy standards that will require cars and light-duty trucks to achieve the equivalent of 54.5 mpg (4.3 L/100 km) by MY 2025. As of December 2012, fuel economy standards were in effect that require MY 2011 to 2016 cars and light-duty trucks to have an average fuel efficiency of 35.5 mpg (6.6 L/100 km) by MY 2016.

THE INDUSTRIAL SECTOR

The industrial sector operates facilities and equipment that are devoted to producing, processing, or assembling goods. As noted in Chapter 1, the major energy uses by this sector are for process heat and for cooling and powering machinery. The sector was the third-largest consumer of energy in 2011, accounting for 21% of primary energy usage. (See Figure 1.10 in Chapter 1.) Petroleum and natural gas were the major energy sources used by the sector in 2011. (See Figure 1.6 in Chapter 1.) Electricity, coal, and renewable sources played smaller roles. The annual energy consumption increased dramatically between 1949 and 1970, but has been relatively flat since then. The industrial sector has reduced its energy usage through efficiency improvements to machines and processes. Another major component of industrial energy conservation has been the onsite production of combined heat and power (CHP).

Combined Heat and Power

Many industrial facilities utilize both heat and electricity in their processes. The electricity is typically purchased from a local utility, while fuel is purchased and burned onsite to produce heat, for example, to boil water to make steam. Great efficiency gains can be achieved by performing both these tasks onsite through a CHP plant. A CHP plant is basically a small power plant that burns fuel to generate electricity; the leftover heat can be used to boil water for steam or for other industrial purposes. In *Combined Heat and Power: A Clean Energy Solution* (August 2012, http://www1.eere.energy.gov/manufacturing/distributedenergy/pdfs/chp_clean_energy_solution.pdf), the DOE and the U.S. Environmental Protection Agency (EPA) indicate that as of 2012 the United States had installed CHP capacity of 82 gigawatts, mostly at manufacturing facilities. According to the agencies, a CHP plant can achieve 65% to 75% efficiency compared with around 45% efficiency for the separate production of electricity and heat.

In August 2012 President Obama issued the executive order "Accelerating Investment in Industrial Energy Efficiency" (http://www.whitehouse.gov/the-press-office/2012/08/30/executive-order-accelerating-investment-industrial-energy-efficiency), which set a national goal of achieving 40 gigawatts of new CHP capacity over the coming decade.

THE RESIDENTIAL AND COMMERCIAL SECTORS

The residential and commercial sectors are often lumped together in discussions of energy conservation because their energy consumption is similar. The primary energy uses for both sectors are for space heating, water heating, air conditioning, lighting, refrigeration, cooking, and running appliances and other equipment.

The two sectors accounted for 11% of primary energy consumption in 2011. (See Figure 1.10 in Chapter 1.) Figure 1.5 and Figure 1.7 in Chapter 1 indicate that the sectors mainly consumed natural gas and electricity in 2011. Petroleum, coal, and renewable energy sources were minor energy sources. As shown in both figures, the overall energy consumption for both sectors increased between 1949 and 2011, especially for electricity. Natural gas consumption has been relatively flat for several decades. Figure 8.4 in Chapter 8 shows that the residential and commercial sectors were the largest users of electricity in 2011, accounting for 73% of total electricity retail sales.

Total energy use in these two sectors has increased over the years because the number of people, households, and offices has increased. In addition, people are using ever larger amounts of electronic gadgets, such as computers, printers, televisions, and copiers. However, increased energy demand has been offset somewhat by efficiency gains in equipment and in building efficiency. In addition, many people have migrated to the South and West, where their combined use of heating and cooling has generally been lower than in other parts of the country.

Efficiency Gains

Energy conservation in buildings in both the residential and commercial sectors has improved considerably since the early 1980s. Among the techniques for reducing energy use are advanced window designs, "daylighting" (letting light in from the outside by adding a skylight or building a large building around an atrium), solar water heating, landscaping, and planting trees. Residential energy conservation is enhanced by building more efficient new housing and appliances, improving energy efficiency in existing housing, and building more multiple-family units.

In *Annual Energy Review 2011*, the EIA indicates that energy consumption per household was 138 million Btu in 1978. Over the following decades it generally declined and was 90 million Btu in 2009. Energy consumption per household has remained fairly steady since 1982. Technology gains have been offset by an increase in the size of new homes and more demand for energy services.

APPLIANCES AND LIGHTING. In 1987 Congress passed the National Appliance Energy Conservation Act, which gave the DOE the authority to formulate minimum efficiency requirements for 13 classes of consumer products. It could also revise and update those standards as technologies and economic conditions change. Energy efficiency has increased for all major household appliances but most dramatically for refrigerators and freezers because of better insulation, motors, and compressors. In addition, efficiency labels are now required on appliances, which makes purchasing efficient models easier for consumers.

The Energy Independence and Security Act of 2007 mandates the gradual phasing out of many incandescent lightbulbs. These bulbs feature a wire filament that generates light and heat when an electric current passes through it. They have been in use for decades and are far less efficient than more modern designs, such as compact fluorescent lightbulbs (CFLs), halogen bulbs, and light-emitting diode bulbs. The DOE provides in "Lighting Facts" (2012, http://www.lightingfacts.com/default.aspx?content/efficiency/summary) the phaseout schedule and lists the types of incandescent bulbs that are affected. In "Compact Fluorescent Light Bulbs (CFLs) and Mercury" (2012, https://www.energystar.gov/index.cfm?c=cfls.pr_cfls_mercury), the DOE notes that replacing just one incandescent lightbulb in every U.S. household with an energy efficient CFL would "save enough energy every year to light 3 million homes and prevent greenhouse gas emissions equivalent to those from about 800,000 cars."

GOVERNMENT INCENTIVES. The EIA indicates in *Direct Federal Financial Interventions and Subsidies in Energy in Fiscal Year 2010* (July 2011, http://www.eia.gov/analysis/requests/subsidy/pdf/subsidy.pdf) that the federal government supplied $6.6 billion in direct expenditures, tax breaks, loan guarantees, and other means of financial support for energy conservation in 2010. This was around 18% of the $37.2 billion that was allocated to the energy industry as a whole that year. According to the EIA, the conservation-related tax measures were mostly directed at residential and commercial taxpayers, for example, to encourage the construction of more efficient homes and buildings and reward the purchase of efficient appliances.

THE DOMESTIC OUTLOOK

In *Annual Energy Outlook 2012*, the EIA notes that U.S. total energy consumption is expected to increase at an average annual rate of 0.3% between 2010 and 2035. Increases in demand are projected to be offset in part by efficiency gains. The EIA predicts that federal and state mandates and incentives will "play a continuing role" in pushing for more efficient technologies and processes. Energy consumption per capita is expected to decrease by 0.6% per year through 2035. In addition, energy consumption per real dollar of GDP is projected to decline by 2.1% per year through 2035. The EIA mentions several factors that it believes will contribute to energy efficiency through 2035:

- Retirement of older less-efficient power plants that are fired by fossil fuels in response to slower growth in electricity demand and increasing environmental regulatory pressure on the electric power sector

- A shift in the industrial sector from energy-intensive manufacturing (e.g., iron and steel) to less energy-intensive manufacturing (e.g., computers and plastics) and service-providing businesses

- Increasing efficiency for freight vehicles, personal vehicles (e.g., automobiles), and household appliances

INTERNATIONAL COMPARISONS OF CONSERVATION EFFORTS

As noted earlier, one indicator of a country's energy efficiency is the amount of energy it consumes per capita (per person). Table 9.1 shows energy consumption per capita for 1980, 1990, 2000, and 2009 for the world, by region, and for the 25 countries with the highest values. Worldwide, energy consumption per capita increased from 63.6 million Btu per person in 1980 to 71.3 million Btu per person in 2009. Between 2000 and 2009 world energy consumption per capita increased by 10%. Asia and Oceania experienced the largest growth on a regional basis; its consumption per capita increased by 52% during this period. The nations with the 25 highest consumption per capita values in 2009 varied greatly in size and

TABLE 9.1

World primary energy consumption per capita, by region and selected country, selected years, 1980–2009

[Million Btus, per person]

	1980	1990	2000	2009	% change between 2000 and 2009
World	**63.56**	**65.62**	**64.87**	**71.26**	**10%**
North America	285.68	276.29	286.03	253.64	−11%
Eurasia	175.75	211.45	135.73	141.89	5%
Europe	135.50	137.03	140.03	134.18	−4%
Middle East	62.24	83.03	100.25	133.33	33%
Central & South America	39.36	40.36	49.32	53.20	8%
Asia & Oceania	19.75	25.17	31.16	47.31	52%
Africa	14.19	15.00	14.98	16.17	8%
Gibraltar	255.94	750.54	3,429.92	1,836.66	−46%
Virgin Islands, U.S.	2,507.34	1,115.27	1,276.12	1,711.16	34%
Qatar	904.62	767.75	1,028.38	1,229.56	20%
Bahrain	395.65	512.03	574.74	764.69	33%
Trinidad and Tobago	159.68	180.25	335.96	697.84	108%
United Arab Emirates	267.23	673.14	579.72	679.38	17%
Iceland	247.59	306.92	424.22	669.25	58%
Netherlands Antilles	1,023.12	782.58	741.87	664.13	−10%
Singapore	182.98	263.40	377.10	485.25	29%
Kuwait	351.08	208.38	460.79	462.28	0.3%
Montserrat	21.75	49.28	189.81	414.97	119%
Norway	328.23	403.31	436.07	407.89	−6%
Canada	394.20	395.06	420.01	389.47	−7%
Luxembourg	389.57	377.88	350.71	352.59	1%
Brunei	442.82	277.82	199.52	324.37	63%
Saudi Arabia	165.97	208.42	227.70	309.30	36%
United States	343.57	338.45	350.19	307.96	−12%
Nauru	246.17	216.41	216.32	275.55	27%
Australia	187.66	218.98	253.66	262.88	4%
Oman	54.55	102.31	140.15	259.13	85%
Belgium	209.63	219.53	266.29	250.17	−6%
Saint Pierre and Miquelon	366.44	567.22	163.57	246.98	51%
Netherlands	226.60	221.44	238.50	242.28	2%
Bahamas, The	285.58	167.93	168.05	241.57	44%
Faroe Islands	152.64	209.64	223.24	237.78	7%

Btu = British thermal unit.

SOURCE: Adapted from "Table. Total Primary Energy Consumption per Capita (Million Btu per Person)," in *International Energy Statistics*, U.S. Energy Information Administration, 2012, http://www.eia.gov/cfapps/ipdbproject/iedindex3.cfm?tid=44&pid=45&aid=2&cid=regions&syid=2005&eyid=2009&unit=QBTU (accessed October 5, 2012)

development status. The United States ranked 17th. It had a better (lower) energy consumption per capita than did the developed nations of Iceland, Singapore, Norway, and Canada. However, the United States had a worse (higher) energy consumption per capita than did Australia, Belgium, and the Netherlands.

Another measure of national energy efficiency is the amount of energy a country consumes for every dollar of goods and services it produces. According to the EIA, in "International Energy Statistics" (2012, http://www.eia.gov/cfapps/ipdbproject/IEDIndex3.cfm?tid=92&pid=46&aid=2), in 2009 the United States lagged behind some industrialized countries in terms of energy efficiency (which is also called energy intensity). The United States consumed 7,340 Btu per dollar (in 2005 U.S. dollars) of GDP, compared with 4,862 Btu per dollar for France, 4,736 Btu per dollar for Germany, 4,675 Btu per dollar for Japan, and 3,898 Btu per dollar for the United Kingdom. That same year, Canada consumed 11,162 Btu per dollar of GDP; Venezuela, 18,182 Btu per dollar of GDP; China, 25,982 Btu per dollar of GDP; and Russia, 31,001 Btu per dollar of GDP.

ENERGY CONSERVATION AND GLOBAL WARMING

As noted earlier, concerns about global warming are driving conservation measures that target fossil fuel combustion. In "What Are Greenhouse Gases and How Much Are Emitted by the United States?" (June 21, 2012, http://www.eia.gov/energy_in_brief/greenhouse_gas.cfm), the EIA indicates that in 2010 about 87% of U.S. greenhouse gas emissions were energy-related emissions. The agency explains that 91% of these emissions were carbon dioxide emissions from fossil fuel combustion. Figure 9.4 and Figure 9.5 provide breakdowns of U.S. energy-related carbon dioxide emissions by major fuel and economic sector, respectively. Petroleum accounted for slightly more of the emissions than did coal or natural gas. The electric power sector contributed the largest share of the emissions, but the transportation sector was a close second. Table 9.2 lists energy-related emissions by source between 1949 and 2011. Note that among petroleum products, motor gasoline has historically contributed the largest share of emissions. In 2011 motor gasoline accounted for 48% of all petroleum emissions.

The U.S. government has focused much of its greenhouse gas reduction efforts on the electric power and transportation sectors. In 2011 they accounted for 73% of the nation's energy-related carbon dioxide emissions. (See Figure 9.5.)

The Electric Power Sector

In March 2012 the EPA (http://epa.gov/carbonpollution standard/actions.html) proposed limiting carbon emissions

FIGURE 9.4

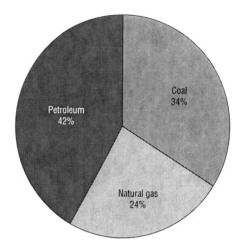

Energy-related carbon dioxide emissions, by major fuel, 2011

Note: Includes small amounts of carbon dioxide from non-biogenic municipal solid waste and geothermal energy (0.2% of total). Totals may not equal sum of components due to independent rounding.

SOURCE: Adapted from "U.S. Energy-Related Carbon Dioxide Emissions by Major Fuel, 2011," in *Energy in Brief: What Are Greenhouse Gases and How Much Are Emitted by the United States?* U.S. Energy Information Administration, June 21, 2012, http://www.eia.gov/energy_in_brief/greenhouse_gas.cfm (accessed October 17, 2012)

FIGURE 9.5

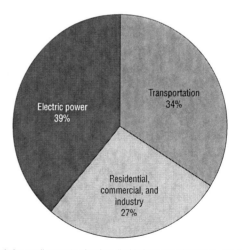

Energy-related carbon dioxide emissions, by sector, 2011

Note: Includes small amounts of carbon dioxide from non-biogenic municipal solid waste and geothermal energy (0.2% of total). Totals may not equal sum of components due to independent rounding.

SOURCE: Adapted from "U.S. Energy-Related Carbon Dioxide Emissions by Sector, 2011," in *Energy in Brief: What Are Greenhouse Gases and How Much Are Emitted by the United States?* U.S. Energy Information Administration, June 21, 2012, http://www.eia.gov/energy_in_brief/greenhouse_gas.cfm (accessed October 17, 2012)

from some newly constructed power plants. As of December 2012, the regulations had not been finalized. The agency

TABLE 9.2

Carbon dioxide emissions from energy consumption, by source, selected years, 1949–2011

[Million metric tons of carbon dioxide[a]]

Year	Coal[c]	Natural gas[d]	Petroleum												Total[b,i]	Biomass[b]				
			Aviation gasoline	Distillate fuel oil[e]	Jet fuel	Kerosene	LPG[f]	Lubricants	Motor gasoline[g]	Petroleum coke	Residual fuel oil	Other[h]	Total		Wood[j]	Waste[k]	Fuel ethanol[l]	Bio-diesel	Total
1949	1,118	270	12	140	NA	42	13	7	329	8	244	25	820	2,207	145	NA	NA	NA	145
1950	1,152	313	14	168	NA	48	16	9	357	8	273	26	918	2,382	147	NA	NA	NA	147
1955	1,038	472	24	247	21	48	27	10	473	13	274	38	1,175	2,685	134	NA	NA	NA	134
1960	915	650	21	291	53	41	42	10	543	29	275	45	1,349	2,914	124	NA	NA	NA	124
1965	1,075	828	15	330	87	40	57	11	627	39	289	65	1,559	3,462	125	NA	NA	NA	125
1970	1,134	1,144	7	394	141	39	78	11	789	41	396	85	1,983	4,261	134	(s)	NA	NA	134
1975	1,181	1,047	5	443	146	24	82	11	911	48	443	97	2,209	4,437	140	(s)	NA	NA	141
1976	1,266	1,068	5	488	144	25	86	13	955	47	506	103	2,372	4,705	161	(s)	NA	NA	161
1977	1,300	1,046	5	520	152	26	85	13	979	52	553	115	2,500	4,846	172	(s)	NA	NA	172
1978	1,298	1,050	5	533	154	26	83	14	1,011	50	544	127	2,548	4,896	191	(s)	NA	NA	191
1979	1,410	1,085	5	514	157	28	95	15	960	48	509	139	2,469	4,964	202	(s)	NA	NA	202
1980	1,436	1,063	4	446	156	24	87	13	900	46	453	142	2,272	4,770	232	(s)	NA	NA	232
1981	1,485	1,036	4	439	147	19	85	13	899	48	376	93	2,122	4,642	234	(s)	(s)	NA	240
1982	1,433	963	3	415	148	19	85	11	892	49	309	80	2,011	4,406	235	5	1	NA	244
1983	1,488	901	3	418	153	19	88	12	904	48	255	98	1,995	4,383	252	7	2	NA	264
1984	1,598	962	3	443	172	17	86	13	914	51	247	106	2,053	4,613	252	10	3	NA	267
1985	1,638	926	3	445	178	17	83	12	930	55	216	93	2,035	4,600	252	13	3	NA	270
1986	1,617	866	4	453	191	15	82	12	958	56	255	98	2,125	4,608	240	14	4	NA	260
1987	1,691	920	3	463	202	14	83	13	982	60	227	106	2,152	4,764	231	16	5	NA	253
1988	1,775	962	3	487	212	14	82	13	1,003	63	249	119	2,246	4,982	242	18	5	NA	266
1989	1,795	1,022	3	491	218	13	82	13	1,000	62	246	118	2,246	5,067	251	19	5	NA	278
1990	1,821	1,025	3	470	223	6	69	13	988	67	220	127	2,187	5,039	208	22	4	NA	237
1991	1,807	1,047	3	454	215	7	71	12	982	66	207	117	2,134	4,996	208	24	5	NA	239
1992	1,822	1,082	3	464	213	6	77	12	999	74	196	135	2,180	5,093	217	26	6	NA	250
1993	1,882	1,110	3	473	215	7	76	12	1,015	76	193	114	2,184	5,185	212	27	7	NA	246
1994	1,893	1,134	3	492	224	7	79	13	1,022	74	183	124	2,221	5,258	218	28	7	NA	255
1995	1,913	1,184	3	498	222	8	78	12	1,044	75	152	114	2,207	5,314	222	29	8	NA	260
1996	1,995	1,205	3	524	232	9	84	12	1,063	78	152	132	2,290	5,501	229	30	6	NA	266
1997	2,040	1,211	3	534	234	10	85	13	1,075	79	142	138	2,313	5,575	222	32	7	NA	259
1998	2,064	1,189	2	538	238	12	75	13	1,107	89	158	125	2,358	5,622	205	30	8	NA	242
1999	2,062	1,192	3	555	245	11	91	14	1,127	93	148	130	2,417	5,682	208	30	8	NA	245
2000	2,155	1,241	3	580	254	10	102	14	1,135	84	163	117	2,461	5,867	212	29	9	NA	248
2001	2,088	1,187	2	598	243	11	92	13	1,151	88	145	132	2,473	5,759	188	27	10	(s)	231
2002	2,095	R1,227	2	587	237	6	98	12	1,183	94	125	127	2,472	R5,806	187	33	12	(s)	235
2003	2,136	1,191	2	610	231	8	95	11	1,188	94	138	140	2,518	5,857	188	36	16	(s)	240
2004	2,160	R1,195	2	632	240	10	98	12	1,214	105	155	142	2,609	5,975	199	36	20	(s)	255
2005	2,182	1,175	2	640	246	10	94	12	1,214	105	164	141	2,628	R5,997	200	35	23	1	261
2006	2,147	R1,158	2	648	240	8	93	11	1,224	104	122	150	2,603	R5,919	R197	37	31	2	R266
2007	2,172	R1,233	2	652	238	5	94	12	1,227	98	129	148	2,603	R6,020	R194	36	39	3	R274
2008	2,139	1,243	2	615	226	2	89	11	1,166	92	111	130	2,444	5,838	R191	37	55	3	289
2009	1,876	R1,222	2	564	204	3	91	10	1,157	87	91	111	2,320	R5,429	R177	40	62	3	R284
2010	R1,988	R1,265	2	R590	R210	3	R94	11	R1,146	77	R96	R120	R2,349	R5,612	186	R43	R73	2	304
2011P	1,874	1,296	2	596	209	2	92	10	1,111	75	86	116	2,299	5,481	186	43	73	8	311

TABLE 9.2

Carbon dioxide emissions from energy consumption, by source, selected years, 1949–2011 [CONTINUED]

[Million metric tons of carbon dioxide[a]]

[a]Metric tons of carbon dioxide can be converted to metric tons of carbon equivalent by multiplying by 12/44.
[b]Carbon dioxide emissions from biomass energy consumption are excluded from total emissions in this table.
[c]Includes coal coke net imports.
[d]Natural gas, excluding supplemental gaseous fuels.
[e]Distillate fuel oil, excluding biodiesel.
[f]Liquefied petroleum gases.
[g]Finished motor gasoline, excluding fuel ethanol.
[h]Aviation gasoline blending components, crude oil, motor gasoline blending components, pentanes plus, petrochemical feedstocks, special naphthas, still gas, unfinished oils, waxes, and miscellaneous petroleum products.
[i]Includes electric power sector use of geothermal energy and non-biomass waste.
[j]Wood and wood-derived fuels.
[k]Municipal solid waste from biogenic sources, landfill gas, sludge waste, agricultural byproducts, and other biomass.
[l]Fuel ethanol minus denaturant.
R = Revised. P = Preliminary. NA = Not available. (s) = Less than 0.5 million metric tons of carbon dioxide.
Notes: Data are estimates for carbon dioxide emissions from energy consumption, including the non-combustion use of fossil fuels. Totals may not equal sum of components due to independent rounding.

SOURCE: "Table 11.1. Carbon Dioxide Emissions from Energy Consumption by Source, Selected Years, 1949–2011 (Million Metric Tons of Carbon Dioxide)," in *Annual Energy Review 2011*, U.S. Energy Information Administration, September 27, 2012, http://www.eia.gov/totalenergy/data/annual/pdf/aer.pdf (accessed October 3, 2012)

indicates in "EPA Fact Sheet: Proposed Carbon Pollution Standard for New Power Plants" (March 27, 2012, http://epa.gov/carbonpollutionstandard/pdfs/20120327factsheet.pdf) that the proposed limit is 1,000 pounds (454 kg) of carbon dioxide per megawatt-hour of gross power produced. The EPA believes that new combined-cycle power plants that use natural gas will be able to meet this limit based on their current designs. Plants that combust coal or petroleum coke will likely require additional technologies, such as carbon capture and storage systems. As explained in Chapter 8, these systems capture carbon dioxide emissions and either sell the gases or sequester them, for example, via long-term storage in deep geological formations.

According to the EPA, as of 2012 California, Oregon, and Washington limited greenhouse gas emissions from some power plants, and other states (e.g., Illinois and Montana) required carbon capture and storage systems for newly constructed coal-fired power plants.

The Transportation Sector

Table 9.3 shows domestic greenhouse gas emissions by transportation mode for 1990 and 2010. In both years highway vehicles accounted for the vast majority of the emissions. In addition, highway vehicle carbon dioxide emissions increased by 24.5% between 1990 and 2010. As described earlier, fuel economy standards were in effect in December 2012 that require MY 2011 to 2016 light-duty vehicles to have an average fuel efficiency of 35.5 mpg (6.6 L/100 km) by MY 2016. These CAFE standards were finalized in 2010 by the NHTSA. At the same time, the EPA issued joint standards that limit carbon dioxide emissions from new vehicles. Table 9.4 shows both sets of standards together. Under these standards MY 2016 automobiles are required to have an average fuel economy of 37.8 mpg (6.2 L/100 km) and emit no more than 225 grams of carbon dioxide per mile.

In 2011 the two agencies finalized the first fuel economy and carbon dioxide emissions limits for non-light-duty highway vehicles. In *EPA and NHTSA Adopt First-Ever Program to Reduce Greenhouse Gas Emissions and Improve Fuel Efficiency of Medium- and Heavy-Duty Vehicles* (August 2011, http://www.epa.gov/otaq/climate/documents/420f11031.pdf), the agencies indicate that the standards apply to certain large pickup trucks, semis (i.e., the tractor component of tractor trailer trucks), buses, and recreational and vocational vehicles. Not all the standards apply to all vehicle types, and the standards vary based on vehicle attributes, such as engine classifications.

The Domestic Outlook

The EIA predicts in *Annual Energy Outlook 2012* domestic energy-related carbon dioxide emissions through 2035 under three scenarios. (See Figure 9.6.) The agency notes that the reference scenario is a "business-as-usual

TABLE 9.3

Transportation greenhouse gas emissions, by mode, 1990 and 2010

[Million metric tonnes of carbon dioxide equivalent]

	Carbon dioxide	Methane	Nitrous oxide
1990			
Highway total	1,190.5	4.2	40.4
Cars, light trucks, motorcycles	952.2	4.0	39.6
Medium & heavy trucks and buses	238.3	0.2	0.8
Water	44.5	0.0	0.6
Air	179.3	0.2	1.7
Rail	38.5	0.1	0.3
Pipeline	36.0	0.0	0.0
Other	0.0	0.2	0.9
Total[a]	**1,489.0**	**4.7**	**43.9**
2010			
Highway total	1,482.5	1.4	16.6
Cars, light trucks, motorcycles	1,077.2	1.3	15.6
Medium & heavy trucks and buses	405.3	0.1	1.0
Water	42.6	0.0	0.6
Air	142.4	0.1	1.3
Rail	43.5	0.1	0.3
Pipeline	38.8	0.0	0.0
Other	0.0	0.3	1.6
Total[a]	**1,750.0**	**1.9**	**20.4**
Percent change 1990–2010			
Highway total	24.5%	−66.7%	−58.9%
Cars, light trucks, motorcycles	13.1%	−67.5%	−60.6%
Medium & heavy trucks and buses	70.1%	−50.0%	25.0%
Water	−4.3%	0.0%	0.0%
Air	−20.6%	−50.0%	−23.5%
Rail	13.0%	0.0%	0.0%
Pipeline	7.8%	0.0%	0.0%
Other	0.0%	0.0%	77.8%
Total[a]	**17.5%**	**−59.6%**	**−53.5%**

[a]The sums of subcategories may not equal due to rounding.
Note: Emissions from U.S. territories, international bunker fuels, and military bunker fuels are not included.

SOURCE: Stacy C. Davis, Susan W. Diegel, and Robert G. Boundy, "Table 11.7. Transportation Greenhouse Gas Emissions by Mode, 1990 and 2010," in *Transportation Energy Data Book: Edition 31*, U.S. Department of Energy, Oak Ridge National Laboratory, July 31, 2012, http://cta.ornl.gov/data/tedb31/Edition31_Full_Doc.pdf (accessed October 3, 2012)

trend estimate, given known technology and technological and demographic trends." The no sunset scenario assumes that certain subsidies and tax credits that are scheduled to expire during the second decade of the 21st century will remain in place through 2035. The extended policies scenario assumes that the government will put into place more extensive subsidies and tax credits and continue to tighten equipment and building efficiency and fuel economy standards. Only the last scenario would actually lower emissions compared with 2012 levels.

The Worldwide Outlook

In "International Energy Statistics," the EIA determines the carbon intensity of countries by comparing the metric tons of carbon dioxide they produce per thousand dollars of GDP. As noted earlier, GDP is a measure of economic activity. Thus, countries with low carbon intensities are emitting less carbon dioxide into the atmosphere per unit of economic activity than are countries with higher carbon intensities. In 2010 the carbon intensity of

TABLE 9.4

Fuel economy and carbon dioxide emissions standards, model years 2012–16

Year	Cars	Light trucks	Combined cars and light trucks
Average required fuel economy (miles per gallon)			
2012	33.3	25.4	29.7
2013	34.2	26.0	30.5
2014	34.9	26.6	31.3
2015	36.2	27.5	32.6
2016	37.8	28.8	34.1
Average projected emissions compliance levels under the footprint-based carbon dioxide standards (grams per mile)			
2012	263	346	295
2013	256	337	286
2014	247	326	276
2015	236	312	263
2016	225	298	250

CO$_2$ = Carbon dioxide.
Note: The required fuel economy, along with projections of CO$_2$ emissions, are shown here.

SOURCE: Stacy C. Davis, Susan W. Diegel, and Robert G. Boundy, "Table 4.19. Fuel Economy and Carbon Dioxide Emissions Standards, MY 2012–2016," in *Transportation Energy Data Book: Edition 31*, U.S. Department of Energy, Oak Ridge National Laboratory, July 31, 2012, http://cta.ornl.gov/data/tedb31/Edition31_Full_Doc.pdf (accessed October 3, 2012)

FIGURE 9.6

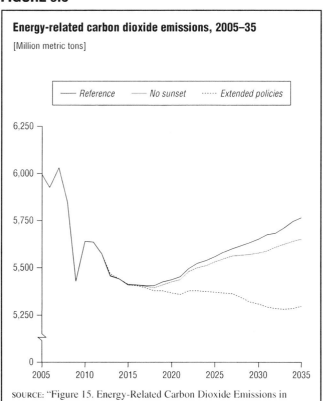

Energy-related carbon dioxide emissions, 2005–35

[Million metric tons]

SOURCE: "Figure 15. Energy-Related Carbon Dioxide Emissions in Three Cases, 2005–2035 (Million Metric Tons)," in *Annual Energy Outlook 2012*, U.S. Energy Information Administration, June 25, 2012, http://www.eia.gov/forecasts/aeo/pdf/0383(2012).pdf (accessed October 3, 2012)

the United States was 0.42. The values for other nations were China, 2.18; Russia, 1.82; Venezuela, 0.93; Canada, 0.46; Germany, 0.27; Japan, 0.25; the United Kingdom, 0.23; and France, 0.18. Therefore, the United States produced less carbon dioxide per thousand dollars of GDP than did China, Russia, Venezuela, and Canada. However, the United States did not perform as well in carbon intensity as did Germany, Japan, the United Kingdom, and France. Each of these nations emitted less carbon dioxide per thousand dollars of GDP than did the United States.

As noted in previous chapters, worldwide energy production is highly dependent on fossil fuel combustion, particularly in developing countries, such as China and India. Therefore, world energy-related carbon dioxide emissions are expected to continue to increase through 2035. (See Figure 9.7.) The EIA predicts that worldwide emissions will grow from 33.3 billion tons (30.2 billion t) in 2008 to 47.6 billion tons (43.2 billion t) by 2035, an increase of 43%. The agency indicates that much of the expected growth is projected to occur in developing countries that are not members of the Organisation for Economic Co-operation and Development (OECD), a collection of dozens of mostly western nations that are devoted to global economic development. The EIA notes that non-OECD emissions were 24% greater in 2008 than OECD emissions. By 2035 non-OECD emissions are projected to be more than twice as great as OECD emissions.

FIGURE 9.7

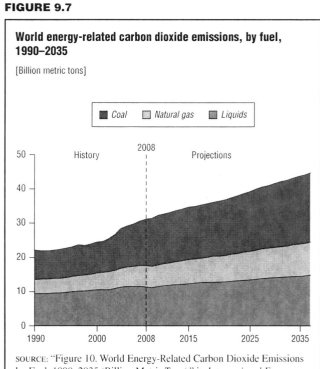

World energy-related carbon dioxide emissions, by fuel, 1990–2035

[Billion metric tons]

SOURCE: "Figure 10. World Energy-Related Carbon Dioxide Emissions by Fuel, 1990–2035 (Billion Metric Tons)," in *International Energy Outlook 2011*, U.S. Energy Information Administration, September 19, 2011, http://www.eia.gov/forecasts/ieo/pdf/0484 (2011).pdf (accessed October 3, 2012)

Another way to look at carbon emissions globally is via carbon intensity, which is measured as emissions per dollar of GDP. Table 9.5 shows U.S. carbon intensity between 1949 and 2011 using real dollars. Domestic carbon intensity fell dramatically from 1,197 in 1949 to 412 in 2011, a 66% decline. In *International Energy Outlook 2011* (September 2011, http://www.eia.gov/forecasts/ieo/pdf/0484(2011).pdf), the EIA predicts that the world's carbon intensity will decline through 2035 "as economies continue to use energy more efficiently."

INTERNATIONAL EFFORTS TO CURB CARBON EMISSIONS. Concern about global warming has spurred many efforts by diplomats and politicians to draw up agreements that would bind countries to certain carbon emissions limits or reductions over time. These efforts date back to 1988, when the United Nations established the Intergovernmental Panel on Climate Change (IPCC), a group of hundreds of the world's leading scientists. Since that time the IPCC has issued several reports (http://www.ipcc.ch/publications_and_data/publications_and_data_reports.htm) on its findings, including the causes and consequences of global warming and climate change. In addition, it has continually urged governments to move quickly with policies to protect the planet. This has proved to be a near impossible task.

In 1997 the United Nations (UN) convened a 160-nation conference on global warming in Kyoto, Japan, to develop a treaty on climate change that would place binding caps on industrial emissions. The resulting agreement, known as the Kyoto Protocol, bound industrialized nations to reducing their emissions of six greenhouse gases by 2012 to below 1990 levels. The United States refused to be a party to the Kyoto Protocol mostly because the protocol did not apply to developing nations such as China. Some of the nations covered by the agreement have found it impossible to meet their limits, sometimes by wide margins. International meetings convened since the late 1990s have failed to find a solution for this problem or to develop new agreements that are acceptable to developed and developing nations alike.

In 2009 a UN Climate Change Conference that was held in Copenhagen, Denmark, did not produce an agreement for the post-2012 period that would bind all nations to specific emissions limits. The Cancun Climate Change Summit in 2010 and a 2011 UN Climate Change Conference in Durban, South Africa, also proved unsuccessful in this regard. However, during the latter conference, the

TABLE 9.5

Carbon dioxide emissions per inflation-adjusted dollar of gross domestic product (GDP), 1949–2011

Year	Metric tons carbon dioxide per million real (2005) dollars
1949	1,197
1950	1,189
1955	1,075
1960	1,030
1965	960
1970	999
1975	910
1976	916
1977	902
1978	863
1979	849
1980	818
1981	776
1982	751
1983	715
1984	702
1985	672
1986	651
1987	652
1988	655
1989	643
1990	628
1991	624
1992	615
1993	609
1994	593
1995	585
1996	584
1997	566
1998	547
1999	528
2000	523
2001	508
2002	503
2003	495
2004	488
2005	475
2006	457
2007	456
2008	444
2009	427
2010	429
2011	412

SOURCE: Adapted from "Table 1.5. Energy Consumption, Expenditures, and Emissions Indicators Estimates, Selected Years, 1949–2011," in *Annual Energy Review 2011*, U.S. Energy Information Administration, September 27, 2012, http://www.eia.gov/totalenergy/data/annual/pdf/aer.pdf (accessed October 3, 2012)

delegates agreed to establish by 2015 a global emissions limit agreement that would take effect in 2020. Negotiations on the future agreement continued at a 2012 conference that was held in Doha, Qatar. As of January 2013, it remained to be seen whether the proposed 2015 agreement would actually come to fruition and if the United States would agree to be bound by it.

IMPORTANT NAMES
AND ADDRESSES

American Gas Association
400 N. Capitol St. NW
Washington, DC 20001
(202) 824-7000
URL: http://www.aga.org/

American Petroleum Institute
1220 L St. NW
Washington, DC 20005-4070
(202) 682-8000
URL: http://www.api.org/

American Wind Energy Association
1501 M St. NW, Ste. 1000
Washington, DC 20005
(202) 383-2500
FAX: (202) 383-2505
E-mail: windmail@awea.org
URL: http://www.awea.org/

Bureau of Land Management
1849 C St. NW, Rm. 5665
Washington, DC 20240
(202) 208-3801
FAX: (202) 208-5242
URL: http://www.blm.gov/

Bureau of Ocean Energy Management
1849 C St. NW
Washington, DC 20240
(202) 208-6474
E-mail: BOEMPublicAffairs@boem.gov
URL: http://www.boem.gov/

Congressional Budget Office
Ford House Office Building, Fourth Floor
Second St. and D St. SW
Washington, DC 20515-6925
(202) 226-2602
E-mail: communications@cbo.gov
URL: http://www.cbo.gov/

Congressional Research Service
Library of Congress
101 Independence Ave. SE
Washington, DC 20540
URL: http://www.loc.gov/crsinfo/

Edison Electric Institute
701 Pennsylvania Ave. NW
Washington, DC 20004-2696
(202) 508-5000
E-mail: electricsolutions@eei.org
URL: http://www.eei.org/

Electric Power Research Institute
3420 Hillview Ave.
Palo Alto, CA 94304
(650) 855-2121
1-800-313-3774
E-mail: askepri@epri.com
URL: http://www.epri.com/

Energy Information Administration
1000 Independence Ave. SW
Washington, DC 20585
(202) 586-8800
E-mail: InfoCtr@eia.gov
URL: http://www.eia.gov/

National Highway Traffic Safety
Administration
1200 New Jersey Ave. SE, West Bldg.
Washington, DC 20590
1-888-327-4236
URL: http://www.nhtsa.gov/

National Mining Association
101 Constitution Ave. NW, Ste. 500 East
Washington, DC 20001
(202) 463-2600
FAX: (202) 463-2666
URL: http://www.nma.org/

Natural Gas Supply Association
1620 Eye St. NW, Ste. 700
Washington, DC 20006
(202) 326-9300
URL: http://www.ngsa.org/

Natural Resources Defense Council
40 W. 20th St.
New York, NY 10011
(212) 727-2700
FAX: (212) 727-1773
E-mail: nrdcinfo@nrdc.org
URL: http://www.nrdc.org/

Nuclear Energy Institute
1201 F St. NW, Ste. 1100
Washington, DC 20004-1218
(202) 739-8000
FAX: (202) 785-4019
URL: http://www.nei.org/

Oak Ridge National Laboratory
PO Box 2008
Oak Ridge, TN 37831
(865) 576-7658
URL: http://ornl.gov/

Organization of the Petroleum
Exporting Countries
Helferstorferstrasse 17
Vienna, Austria A-1010
43-1 21112-3302
URL: http://www.opec.org/opec_web/en//

Solid Waste Association of North America
1100 Wayne Ave., Ste. 700
Silver Spring, MD 20910
1-800-467-9262
FAX: (301) 589-7068
URL: http://www.swana.org/

U.S. Bureau of Reclamation
1849 C St. NW
Washington, DC 20240-0001
(202) 513-0501
FAX: (202) 513-0309
URL: http://www.usbr.gov/

U.S. Department of Energy
1000 Independence Ave. SW
Washington, DC 20585
(202) 586-5000
FAX: (202) 586-4403
E-mail: The.Secretary@hq.doe.gov
URL: http://www.energy.gov/

U.S. Environmental Protection Agency
Ariel Rios Bldg.
1200 Pennsylvania Ave. NW
Washington, DC 20460
(202) 272-0167
URL: http://www.epa.gov/

U.S. Geological Survey
12201 Sunrise Valley Dr.
Reston, VA 20192
(703) 648-5953
URL: http://www.usgs.gov/

U.S. House of Representatives Committee on Natural Resources
1324 Longworth House Office Bldg.

Washington, DC 20515
(202) 225-2761
FAX: (202) 225-5929
URL: http://resourcescommittee.
house.gov/

U.S. Nuclear Regulatory Commission
Washington, DC 20555-0001
(301) 415-7000

1-800-368-5642
URL: http://www.nrc.gov/

U.S. Senate Committee on Energy and Natural Resources
304 Dirksen Senate Bldg.
Washington, DC 20510
(202) 224-4971
FAX: (202) 224-6163
URL: http://energy.senate.gov/

RESOURCES

The U.S. Department of Energy's Energy Information Administration is the major source of energy statistics in the United States. It publishes weekly, monthly, and yearly statistical collections on most types of energy, which are available in libraries and online at http://www.eia.doe.gov/. The *Annual Energy Review* provides a complete statistical overview, and the *Annual Energy Outlook* projects future developments in the field. The website "International Energy Statistics" presents a statistical overview of the world energy situation, and the *International Energy Outlook* forecasts future industry developments. The Energy Information Administration also provides the *Domestic Uranium Production Report*, the *Electric Power Annual*, the *Monthly Energy Review*, and *Natural Gas Monthly*. The *U.S. Crude Oil, Natural Gas, and Natural Gas Liquids Proved Reserves* discusses reserves of coal, oil, and gas. The agency's websites "Energy Explained," "Energy in Brief," and "Today in Energy" contain a wealth of technical information about various energy sources.

The Department of Energy's Oak Ridge National Laboratory publishes the annual *Transportation Energy Data Book* and *Biomass Energy Data Book*. In addition, the Department of Energy makes available information on the development of alternative vehicles and fuels, renewable energy sources, and electric industry restructuring.

The U.S. Environmental Protection Agency maintains websites about hydraulic fracturing and global warming and publishes the annual *Inventory of U.S. Greenhouse Gas Emissions and Sinks*. The U.S. Geological Survey maintains extensive data regarding energy reserves. The U.S. Nuclear Regulatory Commission is also an important source of information and publishes the annual *Information Digest*.

INDEX

"How Can We Compare or Add up Our Energy Consumption?" (EIA), 2

Hydraulic fracturing
   illustrated, 45f
   overview of, 43

Hydrocarbon transformation, 22(f2.1)

Hydroelectric
   consumption/production, 87
   dam, illustrated, 88(f6.4)
   description of, 111
   generators in U.S., 89f
   power, description of, 1
   renewable electricity, hydropower/non-hydropower, 91f

Hydrogen fuel, 99

Hydropower
   environmental issues, 91–92
   market factors, domestic, 90
   outlook, domestic, 90–91
   overview of, 87–89
   production/consumption, domestic, 89
   production/consumption, world, 91

"Hyperion: Committed to South Dakota Refinery" (Dunsmoor), 27

Hyperion Energy Center, 27

I

Imports
   coal, 62
   natural gas, 46
   net natural gas, 49(f3.10)
   oil, 29–31
   petroleum, energy security and, 121
   petroleum, selected countries, 33f, 34t–35t
   primary energy, 13(f1.14)

In situ coal gasification (ISCG), 59

In situ leaching, 72–73

Incandescent lightbulbs, 125

Industrial processing, renewable energy sources, 86

Industrial Revolution, 2, 57

Industrial sector
   energy conservation in, 124–125
   energy consumption in, 6, 9(f1.6)

Intergovernmental Panel on Climate Change (IPCC), 132

Interior region (coal), 58

International Atomic Energy Commission, 78–79

International Energy Outlook 2011 (EIA), 110, 132

"International Energy Statistics" (EIA)
   description of, 109
   on energy efficiency, 127
   on worldwide energy conservation efforts, 130–132

International reserves, 109–110

"Inventory of U.S. Greenhouse Gas Emissions and Sinks: 1990–2010" (EPA), 53

Investor speculation, oil, 38

IPCC (Intergovernmental Panel on Climate Change), 132

ISCG (in situ coal gasification), 59

Isidore, Chris, 25

K

Kerogen
   description of, 21–22
   extraction methods for, 24–25

"Keystone Pipeline: Separating Reality from Rhetoric" (Isidore), 25

Keystone XL project, 25

Kilowatt hour, 71

Klamath River, 92

Krolicki, Kevin, 79

Kyoto Protocol, 132

L

Lease condensate
   description of, 43
   proved reserves, 102f, 106t

Leases, resource, 17

Legislation and international treaties
   Atomic Energy Act of 1954, 17
   Clean Air Act, 68, 94
   Emergency Economic Stabilization Act, 16
   Energy Independence and Security Act, 87, 94, 125
   Energy Policy Act of 2005, 16
   Energy Policy and Conservation Act, 124
   Low-Level Radioactive Waste Policy Amendments Act, 80
   Oil Pollution Act, 39
   Oil Spill Liability Trust Fund, 17, 39
   Public Utilities Regulatory Policies Act, 116

Lenzner, Robert, 38

Liability limits
   energy sector, 17
   Oil Pollution Act, 39

Lighting efficiency standards, 125

"Lighting Facts" (DOE), 125

Lignite
   description of, 57
   production in U.S., 59–60

"Limits of Liability" (U.S. Coast Guard), 39

LLW (low-level waste), 80

Losses, energy
   in biofuels, 93
   electrical, 6
   in energy conversion, 114

"Lower 48 States Shale Plays" (EIA), 52–53

Low-Level Radioactive Waste Policy Amendments Act, 80

Low-level waste (LLW), 80

Low-sulfur coal, 60

M

Market costs, energy, 13–14

Market factors, renewable energy
   federal government support, 86–87
   overview of, 86
   state government support, 87

McGurty, Janet, 94

Measurement
   of electricity, 111
   of natural gas, 42
   of oil, 24

Measuring America: The Decennial Censuses from 1790 to 2000 (U.S. Census Bureau), 6

Middle East
   natural gas production in, 50
   petroleum production/consumption in, 29

Mill tailings, uranium, 80

Mining
   coal, productivity by mining method, 58f
   coal, productivity by region/mining method, 59f
   uranium, 72–73

Mixed oxide (MOX) fuels, 74

"Most Electric Generating Capacity Additions in the Last Decade Were Natural Gas–Fired" (EIA), 113

"The MOX Project" (Duke Energy), 74

N

National Highway Traffic Safety Administration (NHTSA)
   EPA and NHTSA Adopt First-Ever Program to Reduce Greenhouse Gas Emissions and Improve Fuel Efficiency of Medium- and Heavy-Duty Vehicles, 130
   "Obama Administration Finalizes Historic 54.5 mpg Fuel Efficiency Standards," 124

Natural gas
   consumption, domestic, 49–50
   consumption by sector, 50f
   consumption/production, world, 50
   environmental issues, 53
   exploratory/development wells drilled, 108t
   extraction, 42–43
   flow, 51f
   future trends, 52–53
   gross withdrawals, by domestic location, 48(f3.8)
   gross withdrawals, by jurisdiction, 48(f3.9)
   hydraulic fracturing, illustrated, 45f
   imports and exports, 49(f3.11)

proved reserves of crude oil/lease condensate, by state/area, 103*f*

refinery statistics, 26*f*

renewable electricity generation, hydropower/non-hydropower, 91*f*

renewable energy consumption as share of total energy consumption, 87*f*

renewable energy consumption by sector, 88(*f*6.3)

renewable energy consumption by source, 88(*f*6.2)

residential sector energy consumption, 9(*f*1.5)

retail electricity sales, by end-use sector, 115*f*

spent nuclear fuel stored, by state, 81*f*

Strategic Petroleum Reserve, end-of-year stocks in, 36*f*

total electricity and nuclear electricity net generation, 77(*f*5.10)

total energy consumption, by end-use sector, 12(*f*1.11)

total energy supply, disposition, and price summary, 18*t*–19*t*

transportation greenhouse gas emissions, by mode, 129*t*

transportation sector energy consumption, 10(*f*1.8)

uranium drilling activities, 110*t*

uranium production/imports/exports, 75(*f*5.4)

uranium prices, 76(*f*5.7)

world consumption of dry natural gas by region/selected country, 53*t*

world energy-related carbon dioxide emissions, by fuel, 131(*f*9.7)

world installed nuclear power capacity, by country or region, 80*f*

world net electricity generation, by country category, 119(*f*8.10)

world net electricity generation, by country category/energy source, 119*t*

world net renewable electricity generation, by source, 92*t*

world nuclear power units/capacity/production/shutdowns, by nation, 78*t*

world petroleum consumption by region/country, 33*t*

world primary energy consumption, by region and selected country, 20(*t*1.7)

world primary energy consumption per capita, by region/country, 126*t*

world production crude oil by region/country, 32*t*

world production of dry natural gas by region/selected country, 52*t*

Steam engine, 2

Storage

hydroelectric, 87–88

natural gas, underground, 45, 46(*f*3.6)

spent nuclear fuel stored, by state, 81*f*

storage cask containing spent nuclear fuel, 82*f*

Strategic Petroleum Reserve

end-of-year stocks in, 36*f*

overview of, 31–32

"Study Shows Klamath Dam Removal Will Help Farmers, Fish but Skepticism Remains" (Clark), 92

Subbituminous coal

description of, 57

production in U.S., 59–60

Subsidies, 16–17

See also Federal government

Sulfur, 24

Summerland oil field, 24

Supply, energy

total, disposition/price summary, 19(*t*1.4)

total energy, disposition/price summary, 18*t*–19*t*

"Supply of Uranium" (WNA), 110

Sustainability, 2

Swan Hills Synfuel project, 59

Sweet oil, 24

Synfuels, coal, 59

T

Taxes

on coal, 67

targeted, 17

tax provisions, energy related, 16

Technically recoverable resources (TRR)

exploration/development, 105, 106

off-limit areas, 104–105

overview of, 103–104

Tennessee Valley Authority, 116

Tetreault, Steve, 81

Texas, proved reserves of, 102

"Texas Gains the Most in Population since the Census" (U.S. Census Bureau), 6

Thermal energy, 1–2

Thermal generation, electricity, 111

Thermogenic natural gas, 41

Three Gorges Dam, 91

Three Mile Island nuclear accident, 77–78

Tidal power, 99

Tight sand gas, 41–42

Titusville, PA, 21

Transmission

of electricity, 112*f*

natural gas, 45

Transportation

coal, 59

crude oil, 25

crude oil, Keystone XL project and, 25

of natural gas, 43–45

Transportation Energy Data Book, Edition 31 (ORNL), 123

Transportation sector

energy conservation in, 123–124, 130

energy consumption in, 10, 10(*f*1.8)

greenhouse gas emissions, by mode, 129*t*

TRR. *See* Technically recoverable resources

"2011 Domestic Uranium Production Report" (EIA), 72

"2011 Secretarial Determination of the Adequacy of the Nuclear Waste Fund Fee" (DOE), 82

2011 Uranium Marketing Annual Report (EIA), 73

U

Underground storage, natural gas, 45

United Nations, 132

United States

conservation of energy in, 121–122

domestic outlook, 18

energy consumption per capita, 122(*f*9.1)

energy production in, 2, 5–6

natural gas production in, 50

nuclear power production in, 77

Uranium

drilling activities, 110*t*

in electricity generation, 111

fuel rod containing uranium fuel pellets, 76(*f*5.6)

international reserves of, 110

prices of, 76(*f*5.7)

production/imports/exports, 75(*f*5.4)

production/pricing, domestic, 72–75

reserves, 107, 109

UO_2 nuclear fuel fabrication process, 75(*f*5.5)

Uranium 235 (U-235), 71

Uranium concentrate, 72–73

Uranium Location Database Compilation (EPA), 72

U.S. Census Bureau, 6

U.S. Coast Guard, 39

U.S. Crude Oil, Natural Gas, and Natural Gas Liquids Proved Reserves, 2010 (EIA), 101–102

U.S. Department of Energy (DOE)

"Compact Fluorescent Light Bulbs (CFLs) and Mercury," 125

"DOE Sponsored Project Begins Demonstrating CCUS Technology in Alabama," 119

Fuel Economy Guide: Model Year 2012, 16

"Lighting Facts," 125

"2011 Secretarial Determination of the Adequacy of the Nuclear Waste Fund Fee," 82

See also Energy Information Administration

U.S. Environmental Protection Agency (EPA)

biofuel quotas of, 94

CPSIA information can be obtained
at www.ICGtesting.com
Printed in the USA
FFOW041903310513

9 781414 481395